# The Evolved Apprentice

## The Jean Nicod Lectures

Francois Recanati, editor

*The Elm and the Expert: Mentalese and Its Semantics,* Jerry A. Fodor (1994)

*Naturalizing the Mind,* Fred Dretske (1995)

*Strong Feelings: Emotion, Addiction, and Human Behavior,* Jon Elster (1999)

*Knowledge, Possibility, and Consciousness,* John Perry (2001)

*Rationality in Action,* John R. Searle (2001)

*Varieties of Meaning: The 2002 Jean Nicod Lectures,* Ruth Garrett Millikan (2004)

*Sweet Dreams: Philosophical Obstacles to a Science of Consciousness,* Daniel C. Dennett (2005)

*Things and Places: How the Mind Connects with the World,* Zenon W. Pylyshyn (2007)

*Reliable Reasoning: Induction and Statistical Learning Theory,* Gilbert Harman and Sanjeev Kulkarni (2007)

*Origins of Human Communication,* Michael Tomasello (2008)

*The Evolved Apprentice: How Evolution Made Humans Unique,* Kim Sterelny (2012)

# The Evolved Apprentice: How Evolution Made Humans Unique

Kim Sterelny

A Bradford Book
The MIT Press
Cambridge, Massachusetts
London, England

For information about special quantity discounts, please e-mail special_sales@ mitpress.mit.edu

This book was set in Stone Sans and Stone Serif by Graphic Composition, Inc. Printed and bound in the United States of America.

Library of Congress Cataloging-in-Publication Data

Sterelny, Kim.
The evolved apprentice : how evolution made humans unique / Kim Sterelny.
   p.   cm.—(Jean Nicod lectures)
"A Bradford book."
Includes bibliographical references and index.
ISBN 978-0-262-01679-7 (hardcover : alk. paper)
1. Evolutionary psychology. 2. Cooperation. I. Title.
BF698.95.S735   2012
155.7—dc23

2011020798

10  9  8  7  6  5  4  3  2

For my daughter, Kate Nolan Sterelny, with thanks for tolerating my eccentricities and absences

# Contents

# Series Foreword

The Jean Nicod Lectures are delivered annually in Paris by a leading philosopher of mind or philosophically oriented cognitive scientist. The 1993 inaugural lectures marked the centenary of the birth of the French philosopher and logician Jean Nicod (1893–1931). The lectures are sponsored by the Centre National de la Recherche Scientifique (CNRS), in cooperation with the École des Hautes Études en Sciences Sociales (EHESS) and the École Normale Supérieure (ENS). The series hosts the texts of the lectures or the monographs they inspire.

*Jean Nicod Committee*

Jacques Bouveresse, President

Jerome Dokic and Elisabeth Pacherie, Secretary

Francois Recanati, Editor of the Series

Daniel Adler

Jean-Pierre Changeux

Stanislas Dehaene

Emmanuel Dupoux

Jean-Gabriel Ganascia

Pierre Jacob

Philippe de Rouilhan

Dan Sperber

# Preface

In his *Darwinian Populations and Natural Selection*, Peter Godfrey-Smith makes a helpful distinction between philosophy of science and philosophy of nature. The intellectual target of philosophy of science is science itself: most critically, trying to understand how and why science has been such a uniquely successful mode of investigating the world. The intellectual target of philosophy of nature is nature itself; the world in which we live (which, of course, includes humans and their practices, including science). *The Evolved Apprentice* is an essay in the philosophy of nature. It's about nature, because the project is empirical, conjectural and substantive. I aim to develop a plausible, first-approximation model of a striking natural phenomenon: the evolution of the distinctive features of human cognition and human social life. The essay is an essay in philosophy in part because it depends primarily on the cognitive toolbox of philosophers: it is work of synthesis and argument, integrating ideas and suggestions from many distinct research traditions. No one science monopolizes this broad project though many contribute to it. So I exploit and depend on data, but do not provide new data.

It is philosophy, as well, because it addresses one of the core projects of philosophy: human nature and the place of humans in their world. In a remarkably generous review of this book's ancestor, *Thought in a Hostile World*, Alex Rosenberg described my project as that of making a Darwinian world (or worldview) safe for empiricism, of reconciling Locke with Darwin. There is something right about that view of my project, for in explaining most human cognitive competences, I emphasize the power of learning mechanisms rather than preinstalled, competence-specific information. But it is also somewhat misleading, for empiricists have typically been individualists and internalists. I am neither: one message of this book is that human cognitive competence is a collective achievement and a collective legacy; at any one moment of time, we depend on each other, and over time, we stand on the shoulders not of a few giants but of myriads of ordinary agents who have made and passed on intact the informational resources on which human lives depend. Another is that human cognitive competence often depends on epistemic engineering: on organizing our physical environment in ways that enhance our information-processing capacities. The picture of human nature that I develop here therefore has important affinities with some strands of social epistemology and with the work of Andy Clark and his allies. Andy has been a leading light in revealing the many ways in which adaptive thinking depends on an adapted environment.

I will leave a substantive explanation of the ideas of the book to the book itself, but it is worth locating this project in its intellectual market. In the past few years, many versions of evolutionary theories of mind and society have been on offer, so it is worth saying a little about how the apprentice learning model relates to other views under construction. First, my picture is multifactorial and coevolutionary. Evolutionary thinking about humans has been dominated by "key innovation" models: attempts to show that the unique features of human life and mind follow, more or less inevitably, from a single, critical adaptive breakthrough. For example, in the last few years, Richard Wrangham has argued that cooking was the key innovation that drove our increasing differences from other great apes; likewise, Sarah Hrdy has argued that this divergence was driven by reproductive cooperation, especially between females. I am skeptical about all such magic-moment, key-innovation models: I argue instead that coevolutionary, positive feedback loops are responsible for the large and rapid phenotypic divergence between us and our closest living relatives.

The approach that I defend here shares a starting point with mainstream evolutionary psychology. Like the evolutionary psychologists, I am impressed by the informational load on much adaptive human decision making. Many human competences depend on the adept use of often hard-won information. But my picture of how those competences are acquired is much less nativist than that of many evolutionary psychologists. That said, the world does not need another interminable critique of Cosmides, Tooby, and massive modularity, and I do not give one here. I am more interested in developing two other contrasts. Mainstream evolutionary psychology has mostly focused on mapping the end state of hominin cognitive evolution: the project has been to identify the cognitive adaptations of *Homo sapiens* and understand how they operate and interact. I am as interested in the journey as the destination, in understanding the sequence of cognitive, behavioral, and social changes that connect great-ape-like minds, behavioral competences, and social interactions with those of living humans. There is a second contrast. Dan Dennett distinguishes between the identification of the "performance specifications" of a cognitive engine and the cognitive mechanisms that implement the capacities so specified. My main interest is to understand the changes in those specifications over time, and the consequences of those changes. While mechanism constrains capacity and vice versa, I have been as neutral as possible on implementation. So the empirical basis of this project contrasts with that of mainstream empirical psychologists. Given their research focus on cognitive mechanism, their empirical basis is in experimental psychology. Given my focus on capacity, I rely much more heavily on archaeology, paleoanthropology, and ethnography.

Likewise, my approach has many affinities with human behavioral ecology and draws heavily on research in this tradition, especially on work on the economics of foraging, food sharing, and family structure. But there are important differences as well. For one thing, human behavioral ecologists typically work within an individualist paradigm: their models represent agents as maximizing their own fitness. I was once convinced that we could not explain the evolution of the distinctive forms of human cooperation individualistically. I argued that we need the resources of multilevel selection. I am now less sure that this is so, but I certainly do not commit myself to individualism. In my view, it remains likely that selection on groups (especially on culturally defined groups) has been important in human evolution, perhaps especially important in the Holocene. Second, human behavioral ecologists often background an issue I foreground. Adaptive choice in human environments depends on access to relevant information and on the capacity to use that information. No one within the behavioral ecology research tradition would deny that, but their models rarely focus on the informational resources needed for adaptive action. In contrast, much of this book is about access to, and use of, that information. Finally, behavioral ecologists typically treat the environment in which choice takes place as a given: as an external constraint on adaptive choice. In contrast, my book extends niche construction theory to human evolution and explores the ways in which human agents structure and modify their environment. Often we behave adaptively because we have modified our environment in ways that enhance our capacities for adaptive choice. Both individually and collectively, in part we construct our own niche; in part we adapt to our niche.

The apprentice learning model is a model of cultural learning (or social learning; I use these terms interchangeably), so we can make important connections between the picture developed here and the developing theory of cultural evolution. In my view, the expansion of cross-generational cultural learning in the human lineage is a core cause of the increasing phenotypic differences between humans (ancient and modern) and great apes. But I make no commitment here to cultural inheritance as a driver of human evolution, if "cultural inheritance" is understood narrowly. I make no commitment to memes: to discrete, replicated units of cultural information that have fitness interests of their own. I make no commitment to dual inheritance as a general feature of cross-generational cultural transmission. In some contexts, children do learn essentially from their parents, and parents essentially teach only their children. In such circumstances, phenotypic differences within populations can be inherited through cultural transmission, and those traits are subject to selection and evolution. But in many contexts, children learn from other adults of their parents' generation, and many adults influence many children of the next generation. One message of this book is that cross-generational cultural learning can be profoundly consequential to evolution, even when the conditions of dual inheritance are not met. So while influenced by theories of cultural inheritance, models from human behavioral ecology, and the insights of evolutionary psychology, *The Evolved Apprentice* does not neatly fit into any of those traditions. Whether it exemplifies hybrid vigor or mulish sterility I leave for others to judge.

This book has been a long time coming. Some important elements of the picture were already present in *Thought in a Hostile World*; in particular, I argued there for the central role of an organized and adapted learning environment in explaining our cognitive competences. But that book did not come to grips in any serious way with the paleoanthropological and archaeological record of hominin evolution. Nor did I develop there any real picture of how the distinctive features of human cognitive and social life could evolve incrementally. Moreover, the material on cultural evolution was undercooked. So in 2005 and 2006 I began to develop and extend some of the ideas of the 2003 book, especially prompted by my former student Ben Jeffares, who has engaged deeply with these issues himself and who is deeply skeptical about "key-breakthrough" models of hominin evolution. Ben has been a great stimulus and support in the development of much of this material. This process of rethinking and revising was stimulated and enhanced by the Nicod Institute's generous and flattering invitation to present the 2008 Nicod Lectures. To put it bluntly, I did not want to make an idiot of myself, so I was keen to present both a coherent and a genuinely new set of lectures in Paris; I did not want to just rehash and revise earlier ideas. So most of 2008 was devoted to building the first coherent formulation of the basic conceptual model of this book. That model was presented as a set of lectures in Paris in May 2008 and written up as a first draft of a monograph by the end of that year. That draft was inflicted on a large number of friends, colleagues, and slow passersby. The material then slowly fermented for a year as I (like many others in the philosophy of the life sciences) was overwhelmed by the 2009 flood of Darwin celebrations. Thank god he does not turn two hundred every year; I have never been so pleased to see a birthday party finish.

I returned to the manuscript in early 2010 and spent most of that year reworking and revising it. In reworking and revising, I have had a lot of help. It takes a village to raise a book, and I was fortunate enough to live in two villages, one in Wellington, one in Canberra. The graduate student group in the philosophy program at Victoria encouraged me to inflict a full practice run of the lectures on them (and on a few staff interlopers), and their response and feedback helped a lot in the initial formation and clarification of the ideas. I should particularly thank Tony Scott, Kim Shaw-Williams, Alan Poole, Dennis Poole, Matt Gers and David Eng. A second set of responses in Paris was likewise constructive and helpful, especially from Dan Sperber, Hugo Mercier, Brett Calcott, and Peter Godfrey-Smith (who, heroically, read the lectures before their presentation as well as the book manuscript that grew from them). These two audiences saw an early version of the package as a whole; elements of the package were tried out on audiences in Auckland, Canberra, Sydney, Brisbane, Chicago, Oxford, and Vienna. Their responses were almost always helpful and generous. Rachael Brown, Brett Calcott, Steve Downes, Ben Fraser, Don Gardner, Matt Gers,

Peter Godfrey-Smith, Celia Heyes, Peter Hiscock, John Matthewson, Hugo Mercier, Thomas Pradeu, Alex Rosenberg, Nick Shea, Deena Weisberg, Michael Weisberg, and Bill Wimsatt all read the semifinal version, in whole or in part, and forced on me a host of clarifications, amendments, and improvements. Peter Godfrey-Smith, in addition, relentlessly policed my habit of drifting into needless jargon. I fear, though, that some has survived despite his vigilance.

It is always important to acknowledge individual aid, but institutional support should be remembered, too. For much of the gestation of the book, I had two half-time institutional homes: the philosophy programs at Victoria University of Wellington and at the Research School of Social Sciences, Australian National University. They have both been extremely supportive over the last decade, providing a congenial, protected, and often intellectually stimulating environment for the empirically oriented philosophy that I do. Recently that is particularly true of the ANU, with its interdisciplinary Centre for Macroevolution and Macroecology. I was also fortunate enough to be supported by granting agencies in both countries (the Marsden Fund and the Australian Research Council). They supported postdocs, workshops, conference travel, and the like.

Finally, on a personal note, I would like to publicly thank my partner, Melanie Nolan, for her patience and support for my intellectual projects. For much of the time I was at ANU, she had to keep the family home going herself, which she did, with remarkably few complaints (though not zero complaints), despite her own intense research commitments. Even when we have been co-located, I have spent many night and weekend hours in the office. Likewise our daughter, Kate, is tolerant and accepting of both her parents' involvements in research and writing, though she is by no means pleased about my disappearing for days, weeks, and longer to other universities in other cities. I will try to do it less in the future. In the meantime, I dedicate this book to her, as a small token of appreciation of her acceptance of some, at least, of my eccentricities.

Canberra and Wellington, January 2011

# 1  The Challenge of Novelty

## 1.1  Introduction

Hominins split from the chimpanzee lineage six to seven million years ago, and after a relatively unobtrusive beginning, over the last three million years or so, our lineage has diverged sharply from those of our great ape relatives. Hominins became fully bipedal, dependent on technology and cooperation, and, uniquely, combined a fission–fusion social organization with heavy male investment in their offspring.[1] Life history changed: recent hominins live longer than earlier ones and great apes, and our life history has unusual features (adolescents and active postmenopausal females). Our geographic range expanded massively, as did population size. We moved into many new habitats. Our social organization became complex. We acquired novel cognitive capacities: language, metarepresentation, and perhaps even intuitive physics. In sum, change has been rapid: the early molecular clock dates putting the human–chimp split at about six million years ago were greeted with great skepticism because biologists and physical anthropologists did not believe that such great phenotypic divergence could evolve so quickly. Change has been pervasive: morphology, life history, social life, sexual behavior, and foraging patterns have all shifted sharply away from those of other great apes. No other great ape lineage seems to have undergone such a profound transformation: as far as we know, living chimps and gorillas are broadly similar in habitat and ecology to their ancestors of five million years ago. The hominin evolutionary trajectory is bound to be of interest to us, because it is our trajectory. But a disinterested evolutionary biologist would agree that a striking phenomenon exists here, one in need of explanation.[2]

There is a standard picture both of the root cause of this evolutionary trajectory and of the kind of agent that evolved as a consequence of this trajectory. The picture conjoins a selective and an architectural hypothesis. The selective hypothesis supposes that hominin fitness came to depend on managing relations with other hominins. No doubt humans, like other primates, sometimes died of accident, disease, and predation. Our ancestors were not free of the typical burdens of primate flesh and bone. But the distinctive feature—the difference-making feature—of hominin selective environments was the intensity of selection, and the kind of selection, imposed by social interaction. Fitness largely depended on patterns of interaction with other hominins. Those who were more successful in forging cooperative relations, and those who were more adept at interacting with their rivals, left more descendants. Our social environment largely—though not exclusively—shaped our cognitive and behavioral evolution. The demands of an increasingly complex social life required an increasingly sophisticated cognitive response. This basic idea—the Social Intelligence Hypothesis—can be developed in several ways. Robin Dunbar, for example, supposed that increasing group size increases social complexity and puts stress on our memory and conflict management time budgets (Dunbar 2001, 2003). That stress selects for more efficient mapping of the social environment and more effective communication. Geoff Miller's model stresses sexual competition (Miller 1997). But probably the most influential variant of this hypothesis derives from Nick Humphrey's Machiavellian model. According to this model, in hominin social worlds every agent is forced to play social chess, trying to leverage as much profit from social interactions as possible while paying minimal costs. Cooperation with others was an essential ingredient for a successful life, but it had to be carefully managed to secure at least a fair share of cooperation's profit. Clearly, as players become more intelligent, social chess becomes more complex, with selection for still greater intelligence (Humphrey 1976). More on this in the next section.

Hominins are distinctively intelligent, then, largely through selection for social intelligence. This selective hypothesis is conjoined to an architectural hypothesis: the famous modularity model. Notoriously, evolutionary psychologists have developed a modular model of the cognitive engine that has emerged from the complex social worlds of hominin evolution (Carruthers 2006; Pinker 1997; Sperber 1996; Tooby and Cosmides 1992). This picture of our cognitive architecture is motivated by the observation that we solve many day-by-day problems effortlessly, efficiently, and unreflectively. This cognitive efficiency requires special explanation, for we solve many of the routine problems of everyday life only because we are sensitive to many subtle and contextually varying cues. Evolutionary psychologists have followed cognitive psychologists in treating nativist linguistics as a model of the explanation of cognitive competence. Language has seemed a plausible model because nativist evolutionary psychologists take human ancestral environments to pose a set of informationally challenging, recurring, but quasi-independent problems. So, for example, every agent must be able to recognize the norms of his or her social group and to identify acts that would violate those norms. This problem is challenging, as norms are not made obvious by regularities in behavior: you cannot tell what is prohibited just by observing what others happen not to do. There was intense selection on human agents to solve these typical but challenging problems, and as a result we evolved specific adaptations to help us do so. Again, language is a guiding example. Language is a subtle, complex, and abstract communicative system, and our effortless mastery of this system can be explained only by supposing that we come pre-equipped with crucial information about language organization (or so it is often supposed). Other everyday cognitive competences are also informationally demanding. So our effortless mastery of such competences as understanding the minds of others or the norms of our community has a similar explanation. Learning is important. But learning is channeled and shaped by domain-specific, preinstalled information.

I accept with the standard model that many routine human decisions are cognitively demanding, and that managing cooperation has been a crucial driver of hominin evolution. But in the rest of this chapter I argue, critically, that these cognitive demands cannot be managed by prewired modules. Many of the challenges involve evolutionarily novel features of the environment. I also argue, critically, that the standard model misreads the problem of managing cooperation. I argue, positively, that hominins developed a new form of ecological interaction with their environment, cooperative foraging, and this ecological revolution led to positive feedback between ecological cooperation, cultural learning, and environmental change.[3] This feedback dynamic, I argue throughout the whole book, has structured hominin evolution. Our capacity to cope with informationally challenging problems in novel environments (some of our own making) depends on this dynamic. In particular, it depends on the construction of minds and social environments adapted to efficient, high-volume social learning.

In my view, the standard picture understates the dynamism and connectedness of hominin evolutionary environments. As a consequence, it mischaracterizes the information-using preconditions of a successful hominin life. The questions that hominin environments asked of our ancestors are not quasi-independent. Hominins evolved into social, cooperative foragers. As a result of that economic transformation, foraging practice, technology, social organization, and human demography all interact. For example, the "broad-spectrum revolution" names one striking episode in the human past as our ancestors shifted from relying heavily on hunting large and medium-size herbivores to using a much more extensive range of animal, plant, and marine resources (Stiner 2001; Stiner and Kuhn 2006). This economic transition changed human group size, social organization, technology, and foraging practice. Specialization, differentiation, and population density all increase together. Changes in any one of these variables affect the others.

Moreover, change was pervasive. Hominins evolved in times of increasing climatic variability, and by about two million years ago they had spread far and wide from their original East African epicenter. So the physical environments of our ancestors became more variable and heterogeneous (Finlayson 2009; Potts 1996). Furthermore, and most importantly, hominins became increasingly potent ecological engineers. The hominin footprint on the local environment became ever more marked and more pervasive. Thus by a half-million years ago (perhaps earlier) our ancestors had become effective predators of medium to large herbivores (Foley and Gamble 2009; Jones 2007). This affected how hominins experienced their environment and how selection acted on our ancestors. For example, as weapons, ecological expertise, and cooperation improved, the impact of many predators would decline. But hominin activities also reshaped the environment itself. Very likely, through competition and active persecution, those hunters were affecting the absolute numbers, distribution, and behavior of rival predators. Thus the environments of hominin evolution have been unstable both physically and biologically.

They have also been unstable socially. Group size, the extent and nature of the division of labor, the extent of social hierarchy, and the importance and nature of interactions with other groups all affect an agent's social world. None of these factors has been constant over the last hundred thousand years. Robert Foley, in particular, has long emphasized the relationship between resource distribution and sexual dynamics. For example, if crucial resources are clumped in rich, predictable local patches (like a salmon run), then one or a few males, by seizing control of resources, can thereby ensure sexual access to the women needing those resources. This resource distribution opens the door to polygyny. In contrast, if resources are pepper-potted unpredictably throughout the environment, women will scatter, chasing those resources, closing the door to polygynous strategies. In my view, human worlds have been heterogeneous psychologically as well as socially and physically: the psychology of other agents has also varied over the last hundred thousand years. The standard model rules this possibility out. If our minds are (mostly) ensembles of (largely) prewired modules, then human nature is largely the same everywhere and when. But we are pervasively and profoundly phenotypically plastic: our minds develop differently in different environments. The extent and nature of this plasticity is controversial, but its existence is not. Humans obviously differ in skills, capacities, and information, and those differences are relevant to social chess. Likewise we clearly have some emotional and motivational complexes whose development is channeled by specific cultures; the response to a perceived insult is very different in a "culture of honour" (Nisbett and Cohen 1996) than in, say, a culture like that of the Faroe Islands, which values social harmony and peacemaking highly (Gaffin 1995). If motivational, emotional, and decision-making mechanisms are indeed plastic in important ways, important differences in human socio-foraging worlds will result in importantly different inhabitants of those worlds.

A central dynamic of human evolution, then, is that while the informational demands on adaptive human action have long been significant, these informational prerequisites are neither stable nor relatively discrete. The standard model is right to insist that many everyday challenges of human social life impose high cognitive loads, and that our response to these challenges is typically competent. Such ubiquitous competence does indeed require special explanation. How is it, for example, that almost all of us master and respond to the norms of our immediate circle? In many cases, this competence does not depend on our being prewired with most of the crucial information needed for adaptive response. We sometimes blunder; we sometimes overlook the obvious despite repeated exposure. But we often respond competently to novel high-cognitive-load problems.[4] The standard model overstates the informational independence and stability of the challenges we usually meet. As I see it, then, a key challenge to models of human evolution is to explain competent response to novel problems. The standard model recognizes the centrality of learning to human social life, and different versions of this model vary in the extent to which they appeal to prewired capacities. But to the extent that those models explain competence by appealing to preinstalled information, they are not well designed to explain competence in the face of the new.

## 1.2 The Social Intelligence Hypothesis

Hominin cognitive evolution cannot have been driven mostly by external environmental change, as then we would expect similar trajectories in other species, and that we do not see. Five million years ago, our ancestors were unobtrusive elements of the East African mammalian fauna. We now inhabit essentially every terrestrial habitat, in numbers unprecedented for a large mammal, and we have transformed most of the world's ecologies. The speed and extent of this evolutionary transformation suggest that it has been driven by a positive feedback loop. It is unique; no other great ape lineage is a mirror site, reflecting a similar response to external events. This suggests that the dynamics are internal, though presumably triggered by some idiosyncratic feature of our early history.

According to the standard model, the feedback loop derives from the problem of managing cooperation, a problem that becomes ever more crucial, and ever more difficult, as human agents become more intelligent. As the standard model represents the problem of cooperation, it rests on the strategic aim of enjoying the benefits of cooperation without being exploited by others. Cooperation can be highly profitable, because a group acting jointly can generate a higher return than the sum of each of them acting individually. Collective defense, for example, will typically be far more effective than individual defense. Hominin evolution, among much else, is one long lesson in the profit of cooperation and the power over the world that derives from successful cooperation at and across generations. So cooperation has a potential benefit, but only if the costs of defection can be contained. Cooperative actions are not free, and the benefits of cooperation often do not fully depend on every agent paying the full cooperation cost. Collective defense can still be successful even if one defender lurks in the rear. These circumstances generate a temptation to avoid the costs of cooperation while collecting the benefits. Thus it seems that in many circumstances, even when cooperation is profitable, it will not evolve. Cooperation cannot persist if free riding is still more profitable. So cooperation will not establish if it is too difficult or too expensive for cooperators to exclude free riders.

This analysis of the "hard problem" of cooperation is reflected in the traditions of both evolutionary models and experimental economics. Much evolutionary modeling of cooperation is based on variations of iterated prisoner's dilemma themes (Hammerstein 2003). In these models, the rewards of successful cooperation (and those of defection and of trust betrayed) are free parameters to be adjusted as the modeler chooses. The models explore the consequences of different patterns of interaction, the effects of punishment, of error, of group structure, of the effects of the manipulation of rewards and costs. They do not explore the mechanisms that generate the rewards of cooperation. The same is true of experimental economics. For example, in many public-goods games, the central pool that is the reward of cooperation is simply by experimental fiat double the total of the individual contributions. The experimental subjects need to commit to cooperation. But that cooperation involves no collective action or joint problem solving. Rather, these experiments investigate the conditions under which cooperation stabilizes or decays, conditional on the ways the profit of cooperation is divided among the players.

Machiavellian hypotheses thus focus on this cognitive challenge of managing cooperation in an environment in which defection is a threat rather than the problem of coordination, of organizing collective action so that it generates a cooperation profit. Cooperation is so profitable that it eventually became an obligate feature of hominin lifeways. Going it alone has probably not been an option for hundreds of thousands of years. But in such environments, agents must calculate and police reciprocal bargains, scrutinize signals for honesty, decide on disclosure principles, negotiate alliances, decide whether to defect. As other agents become more intelligent, these decisions become more demanding. As cognitive sophistication increases, social environments become more demanding. This selects for further cognitive complexity. Perhaps an initial shift toward cooperation had some local external cause. It may be that early hominins—australopithecines of three or four million years ago—had to band together cooperatively for protection as their forests turned into woodlands and savannas in a warmer and drier world. At this stage of hominin evolution, cooperation was driven by external, environmental factors. But once cooperation and its management became central to the lives of our ancestors, that established a positive feedback loop between social complexity and individual cognitive capacity.

There is a natural link between the Machiavellian version of the Social Intelligence Hypothesis and a modularity hypothesis, for most candidate modules are tools for social life. If prudent cooperation was central to a successful hominin life, and prudent cooperation was stable only through vigilant mutual scrutiny, we might well expect special adaptations to monitor social exchange and to monitor norms and norm violation. Most obviously, folk psychology will be a crucial resource in cooperation management. It is essential to track the beliefs, preferences, and intentions of others in a world in which partners are necessary, but in which they are at best reluctantly honest and kept so only by sleepless vigilance. Machiavellian versions of the Social Intelligence Hypothesis predict that we have minds with a suite of adaptations for a social life revolving around bargaining, exchange, and honesty assessment. That is just the kind of mind that advocates of massive modularity hypotheses think we have.

Moreover, the cognitive complexity of other agents, and the social complexity that cognitive complexity generates, explain why routine human decision making has a high cognitive load, and hence why everyday competences need to be supported by special tools. We are individually complex agents living in, and contributing to, socially complex worlds. The factors that ramp up the informational demands on routine decisions include the following:

(i)   We have many needs, so trade and exchange are complex, with multiple trade-offs.

(ii)   Many human groups exhibit significant division of labor and specialization, so many humans have lived in groups with significant horizontal differentiation.

(iii)   Many human groups, including some foraging cultures, exhibit significant vertical complexity: individuals form parts of families, which in turn are components of bands, totem groups, and the like. Thus even small-scale traditional societies often have layers of social organization between individual agents and the group as a whole.

(iv)   We are long-lived, with good memories, and form long-lasting, high-stakes relationships. Entering into a sexual or social alliance is often a high-risk, high-reward decision.

(v)   If Robin Dunbar is right, hominin social worlds have trended up in size. Hominins, increasingly, have needed to keep track of more agents and to map their social relations.

(vi)   Sex is complex, as we are social, quasi-monogamous primates with male investment and somewhat concealed female ovulation. Moreover, we live in a fission–fusion society with a sexual division of labor. Males cannot guarantee paternity by direct vigilance of female behavior. Sex is further complicated by partial reproductive cooperation between relatives; for example, by a grandmother caring for her daughter's children.

(vii)   We pool information, as well as cooperating to make direct economic gains. So communication plays an especially important role in human social worlds. But not all communication is honest.

(viii)   Human social worlds are organized around norms, not just regularities or habits. Violations of norms are often punished, formally or informally, even when the norms are not made explicit.

(ix)   Agents are only partially transparent to one another. We signal richly, but some of those signaling systems are arbitrary, referential systems with low intrinsic reliability. We have considerable voluntary control over facial expression, stance, and voice, and so we can partially fake and suppress many natural cues. We have stealth and deception capacities.

Ordinary human decision making, then, takes place in a translucent social world. Often relevant information is available, information that would guide adaptive decision making were an agent aware of it and able to assess its relevance and reliability. But cues are often not perceptually salient. Their relevance is often not obvious, and their reliability is difficult to assess. Our social world is translucent because it is the result of a Machiavellian evolutionary dynamic.

The cognitive challenges of policing the division of collective and cooperative products are real. The problems of deception and defection are not just artifacts of contemporary mass societies. They existed in the social worlds in which the cooperative framework of human life evolved. But defection management is not all that is needed to keep cooperation stable. To be stable, it must also be profitable, and profitable cooperation often requires coordination, and that is often cognitively demanding. Indeed, in small-scale foraging social worlds, the cognitive problem of effective coordination is *more demanding* than that of detecting defection. The standard picture is right to identify the evolution of stabilized, extensive, obligate cooperation as the core, distinctive feature of hominin selective environments. But that picture misrepresents the task demands on cooperation, for it focuses on explaining how the profit of cooperation is distributed in ways that do not undermine the motivation to cooperate. I suggest (following Calcott 2008b) that a prior question is equally pressing: how does hominin cooperation generate a profit? I begin to answer that question in the next section, and in doing so, I develop the idea that the task demands facing hominins were not just variable over time; they were interdependent.

## 1.3  Cooperative Foraging

Hominin social complexity has certainly increased. But there has also been a transformation in the ways that hominins interact with, and extract resources from, their environment. The (gracile) australopithecines and early *Homo* were, as far as we can tell, generalist scroungers, subsisting on the proverbial nuts and berries, with the odd grub, slow lizard, and scavenged carcass fragment thrown in. By two hundred thousand years ago, perhaps much earlier, our ancestors were dominating predators.[5] In sharp contrast to other predators, those hominins often specialized in the prime adults of their target species, typically large ungulates (Stiner 2002). Hominins went from being food to taking food from other members of the predator guild. The shift from marginal scrounging to major predator status most likely took place via increasingly aggressive scavenging. Thrown volleys of rocks would be no great threat to, say, a mobile leopard. But they would genuinely endanger one immobilized by the need to defend a kill. Importantly, the shift to predation preceded the invention of long-distance lethal weapons. We killed large animals before the invention of spear-throwers; bow-and-arrow technology, or poison-tipped weapons (Marlowe 2005). Spears (and perhaps killer frisbees) sufficed.

Later still, probably as a result of living in larger groups and of our increasing ecological footprint, the range of resources that humans harvested expanded greatly. For example, in Europe, by the time the Ice Age was ending, this shift intensified (Stiner 2001). The systematic exploitation of plant-based resources increased. Fish and other marine and riverine resources became important. Waterfowl and smaller game were taken with specialist equipment. Indeed, in general, this expansion of the resource base is paired with an expansion of specialized toolkits. Foragers developed specialized toolkits and techniques to grind grain and make bread, to harvest water-based resources, and to catch smaller game economically.

I take these shifts in ecological role to be a clear historical signal of the invention and establishment of a new lifeway, built around a new mode of foraging. By two hundred thousand years ago, and most probably much earlier, hominins had evolved into social foragers.[6] Such foragers depend on harvesting high-value but heavily defended resources. The regular exploitation of those resources (at tolerable risk loads) depends on some mix of (i) rich, targeted ecological information (so, for example, tubers are a rich carbohydrate store, but they must be found, recognized, detoxified, processed); (ii) cooperation; and (iii) technology. Typically all are needed, though the exact mix will vary with time, place, and target. Hunting large animals is intrinsically risky, and it required technology to be integrated with a detailed understanding of the prey, its capacities, habits, and likely reactions, and to skilled, coordinated group hunting. Truly lethal weapons are needed before individuals and small groups can take large and dangerous prey. Neanderthals, like sapiens, were effective hunters of medium to large prey (d'Errico and Stringer 2011), and their use of heavy spears is known from the archaeological record of about 400,000 years ago (Thieme 1997). So it is likely that the common ancestor of our species and the Neanderthals was also an effective social forager, though it is possible that the sapiens and Neanderthal lineages evolved in parallel from a less adept ancestor.

On this view of hominin evolution, as with Machiavellian models, cooperation is central to our evolving cognitive capacity. But our conception of the informational challenge changes. Cooperative foraging (and especially cooperative hunting and cooperative defense against predation) requires coordination, and thus communication. Cooperative hunters must plan and coordinate before targeting potentially difficult and dangerous targets, especially if there is task specialization. But even if there has been advanced and expert planning, on occasion not everything will go according to plan. Agents will have to react on the fly, sometimes in novel situations, and often with imperfect information. They will make high-stakes decisions under time pressure, based on their reading of the physical and biological context and on their expectations of others' reactions, and with rather limited prospects for communication and consultation. No doubt those fast-response decisions often failed. But they were good enough, often enough, for lifeways based on cooperative foraging to establish and spread, and that is impressive in itself, for these are high-load, high-stakes decisions. Hunting and killing large animals with a sharp stick is no easy project. Both the risks and the rewards are great. These are large, concentrated resource packages. But catastrophe is just a moment's inattention away.

Hominin life, then, came to depend on the rich resources that cooperative foraging delivers. In turn, cooperative foraging demands care, coordination, and skill. In the next section, I explore one aspect of skill: its dependence on social learning.

## 1.4 Cooperative Foraging and Knowledge Accumulation

Cooperative foraging is one key transition in hominin evolution. Such foraging is a profitable but demanding form of life, for the profit of joint foraging typically depends on effective coordination, often in far from ideal circumstances. Agents must often make decisions when distracted, under time pressure, in danger, and with obscured lines of sight, often in noisy or confused surrounds. The decision-making environment is at best informationally translucent. I suggested in the previous section that hunting large game cooperatively with limited technology depends on effective coordination and the use of transient target information. But it also depends on a rich understanding of stable features of the physical, biological, and technological environment. For example, Kim Shaw Williams (2011) shows that tracks, scats, browse marks, and other physical signs of passage are rich with information for the prepared mind and eye. If the surface is suitable, a tracker can read information about the identity, direction, and time of travel of local fauna, and there is as well much information to be had about the interactions among them (this information is beautifully illustrated, in a European context, in Ennion and Tinbergen 1967). Foragers do have prepared eyes and minds; they are expert natural historians of their local patch. Typically much of this information is acquired culturally; for example, a striking set of photos in Morrison 1981 shows Aboriginal children learning how to recognize different tracks by being shown how to reproduce them. So a second cognitive precondition of cooperative foraging is the existence of a flow of information across the generations. Cultural learning of this kind can begin and can become important without the active cooperation of the source of information. Agents leak information in their everyday activities. Moreover, they often adaptively structure the learning environment of their young as a by-product of their own utilitarian activities. However, I argue that high-volume, high-fidelity cultural learning depends on informational cooperation between source and sink (the target to whom information flows) in an appropriately organized environment. It depends as well on specific perceptual and cognitive adaptations, probably of the source as well as the sink.

Sometime between about 120,000 and 50,000 years ago, human cultures began to resemble those known from the ethnographic record. By 50,000 years ago, humans had diverse toolkits: they exploited a wide range of materials in their technology, made complex tools, exploited many natural resources for food, buried their dead, had distinct local styles, and made objects that appear to be decorative rather than directly utilitarian. In the jargon of archaeology, they were "behaviorally modern." Behaviorally modern human cultural worlds depend on high-volume, high-fidelity cultural learning. The elaboration of technology (and thus of technique) depends on a group being able to retain the cognitive capital it inherits, occasionally adding an innovation to it, then transmitting that enhanced capital to the next generation with high fidelity. Indeed, it is arguable that behaviorally modern humans differ from their predecessors just through the establishment of social environments in which high-volume, high-fidelity social learning is robust (more on this in chapter 3). But earlier humans also depended on quite high-volume, high-fidelity social learning. Exploiting high-risk, high-return resources is itself a signature of the capacity to preserve and transmit informational resources. The Neanderthals who regularly exploited elk and other large European ungulates, and the Middle Stone Age Africans who specialized in similar targets, were skilled and knowledgeable. So, for example, Neanderthals brewed adhesives for their hafted weapons out of birch bark, using processes that depend on a surprisingly precise control of temperature (Nowell, forthcoming; d'Errico and Stringer 2011). Expertise and cooperation compensated for limited technology. Likewise the ancient tuber and corm harvesters depended on much hard-won information, if underground storage organs really were important resources from *erectus* on. Fruits are designed to be eaten. But plants do not welcome herbivore consumption of their storage organs, and hence they are protected both mechanically and chemically. It takes a well-informed mind to find these organs, extract them, and make them edible by soaking, cooking, and the like.[7]

The idea, then, is that positive feedback links social foraging and intergenerational social learning. Intergenerational learning provides much of the informational fuel that makes social foraging successful, and the rewards of social foraging support the life spans and expensive metabolisms that make extensive intergenerational learning possible. A skeptic might concede that once humans are behaviorally modern, high-fidelity, large-bandwidth, cross-generational social learning plays a central role in human lives and societies. However, this is a relatively recent feature of hominin history (the challenge runs), and hence social learning in positive feedback with collaborative foraging cannot be a crucial driver of hominin evolution. Cross-generational human learning became a central part of human lives only after human minds, bodies, and social worlds had largely evolved.[8] In support of this deflationary view of cross-generational social learning, a skeptic might legitimately point out that until the last hundred thousand years or so, hominin technology seems to have been very conservative, with a small, slowly growing toolkit. Hominins may have innovated, but if they did, those innovations rarely became the new standard practice within the innovator's community. In a world in which children reliably absorbed parental lore, we should expect to see a less-conservative material culture, for children would inherit an information platform from their parents as a basis for further improvement

Unsurprisingly, I disagree. The capacity *to add* regularly to cognitive capital by reliably preserving and amplifying innovation may be relatively recent. Even so, the reliable *preservation* of expertise is ancient. Around 400,000 years ago, humans were using material technology that could not have been reinvented anew each generation, helped only by some minimal prompting by the elders. Making stone spearheads is a difficult art, with multiple processing stages. Control of shape is increasingly difficult as the target to be made becomes smaller, if it is symmetrical, and if one dimension is elongated. Yet the technology 300,000 years ago—so-called mode 3 technology— required close control of shape in working relatively small artifacts (Foley and Lahr 2011).[9] Likewise, the natural-history expertise essential to cooperative foraging could not be rebuilt every generation without substantial informational input from the previous generation.

Both hominin minds and hominin social environments are adapted to the social acquisition, use, and transmission of ecological and technological expertise. Without such adaptations of minds and social environments, life as a social forager could not have evolved.

Social foraging, then, is informationally demanding over short time frames through the requirements of joint and coordinated action. It is informationally demanding over longer time frames because it rests on a reservoir of skill and expertise. Moreover, social foraging requires the *integration* of ecological, technological, and social information. A group of foragers putting a hunting party together needs to appreciate both their local ecology and the capacities of their partners. The same is true of on-the-fly decision making. For example, effectively responding to an emergency requires agents to integrate what they know of the situation—the level and nature of the threat, the lay of the land, the potential responses—with their knowledge of their social partners. Agents responding to a threatened attack need to understand who stays calm, who panics, who is a hothead. Moreover, they need to factor in the physical condition of their partners. The right response to injury, fire, or flood depends on specific local circumstances and on the abilities and frailties of those who face emergency. Response cannot be too stereotyped. No doubt social foragers quite often made poor decisions in response to crisis. But the world of 150,000 years ago was much more dangerous than it is now (setting aside the danger posed by humans), and so the persistence of this lifeway in that dangerous world shows that social foragers often responded appropriately to the unexpected. The basic structure of human life history, with its extended periods of juvenile dependence, requires that on average, mortality is low once adulthood is reached (as we will see in sec. 4.3). The evolution of human life history patterns is hard to date, but researchers generally agree that by the time of the erectines (around 1.8 million years ago), a significant shift from ape to human life history patterns had occurred. This extension of life history (humans live a good twenty years or more longer than chimps) shows that while social foragers had many dangers to negotiate, they usually did so successfully. The assembly, integration, and intergenerational transmission of social, ecological, and technological information have deep roots.

A quick review of progress to date is in order. Like others, I think the expansion of cognition in the hominin lineage is intimately connected to the expansion of cooperation in that lineage. But in contrast to the usual defection-management perspective, I see the key cognitive preconditions of cooperation as being those necessary for coordination and effective joint action. In a complex and risky environment, profitable cooperation depends on more than trust, on more than the confidence that you will not be ripped off. It also depends on teamwork, with a mutual awareness of one another's abilities, and on integrating this social information with appropriate information about the world: the target of joint action. Once these conditions were met, joint action was profitable. Indeed, it became increasingly profitable as target information and technology were harnessed to improved coordination. The evolution of coordinated action depended on improving capacities to coordinate and on improving, harnessing, and retaining for the next generation information about foraging targets and techniques. Critically, once this form of cooperation is established, it triggers a positive feedback loop between environmental change, ecological cooperation, and cross-generational learning. Cooperation increases the pace of environmental change, and the nativist solution to the problem of high cognitive load is increasingly restricted to special cases.

## 1.5   Life in a Changing World

Hominins have not evolved in a stable world. As Rick Potts (1996, 1998) and Clive Finlayson (2009) emphasize, the world of hominin evolution has been increasingly climatically unstable: the Holocene is an aberrant stretch of stability against a shifting background. Likewise Pete Richerson has recently argued that much of human evolution should be understood as a response to climate, both to its increasing instability before the Holocene and to its anomalous stability since (Richerson, forthcoming; Richerson and Boyd 2002). But more important still, cooperative foraging is such a powerful mode of interacting with the environment that it directly and indirectly transformed the hominin environment and thus the ways in which selection acted on our ancestors.

Cooperation (perhaps in conjunction with other adaptations) has allowed the hominin lineage to penetrate new regions and habitats. Hominin environments have become increasingly variable across space as hominins have become increasingly widespread ecologically and geographically. Moreover, cooperative foraging has an increasingly heavy ecological and physical footprint over time. The populations of target species are depleted.

Predators become increasingly rare, wary, or both. These environmental effects also create coevolutionary opportunities for species that will eventually domesticate, and for scavengers of various sizes (rats, mice, cockroaches, lice). We experience new pathogens as we change our mobility, residence patterns, and population size. Landscapes are altered. Australian Aborigines, for example, have long used fire as a tool both to clear underbrush, making game more accessible, and to induce a flush of grass growth, encouraging their target species to return to their hunting grounds. These altered fire regimes, with much more frequent burning, have had dramatic long-term effects on Australian landscapes (Bliege Bird et al. 2008; Pyne 1998).

So the direct effect of social foraging is significant and cumulative as environmental change becomes more rapid and intense. That was especially true once innovations were more reliably preserved, transmitted, and built on, for then individual and collective effects on environments increase. Consider again the elaboration of the control of fire, from true hearths and earth ovens to container-based cooking, pottery, and other technology that depends on the control of heat. These developments increasingly altered humans' experience of their environments. The same is true of clothing (Gilligan 2007), shelters, watercraft, and tools and weapons. But the evolution of social foraging had profound effects on the social environment as well, by both selecting for and making possible (through an increased period of juvenile dependence) increases in the fidelity and volume of cultural learning (Carey and Judge 2001; Kaplan et al. 2000; Robson and Kaplan 2003). Elaborated social learning almost certainly evolved because of selection for utilitarian expertise, and that in itself would change social life as children became more interested in adult activities, and adults more tolerant and communicative. Indeed, it has even been suggested that elaborated social learning has led to a distinct form of social hierarchy based on esteem and prestige. Esteem and deference are the price the less able pay to experts for access to their expertise (Henrich and Gil-White 2001). But once evolved, those capacities changed social life in even more profound ways as they were exapted for other purposes, including machinery for social cohesion. Once social learning became central to human psychology and social life, other important changes in human social life followed, as groups began to consciously and publically identify themselves as groups. Public symbols in various forms—song, ritual, physical symbols, public art—are part of the machinery through which group cohesion and identity emerge. Mere regularities become entrenched as customs and norms; they become markers of who we are, not just of what we happen to do. These symbol systems depend both directly and indirectly (via the technology needed to make them) on elaborated cultural learning. Music and ritual, too, are transmitted socially, but once established, they profoundly change social lives (Mithen 2005).

The effects of social foraging on demography and group size also increase the pace and intensity of environmental change. All else being equal, improving the efficiency with which humans extract resources from their environment will result in an expanding population and an increase in group size. Larger groups preserve informational resources more reliably, for learners have more expert models from whom they can pump information, and expertise is less likely to be lost by unlucky accident. But as Haim Ofek argues in *Second Nature*, size makes the benefits of specialization more available. There is a market for special skills, so larger groups can divide labor more finely. Ofek conjectures that fire keeping was the first form of labor specialization. If he is right, that specialization preceded behavioral modernity. But as the returns of social foraging increased, especially after behaviorally modern humans began to depend on the efficient harvesting of many different resources, among larger groups there would have been important incentives for specialization. If specialists are more likely to successfully innovate in their field of specialization, as seems likely, positive connections will develop between elaborating social foraging, increased group size, and the rate of innovation.

In sum, feedback loops form between individual cognitive capacity, social organization, and the pace of environmental change. Environmental change, in turn, is important because the informational requirements on adaptive action vary as the environment varies. The right choice often depends on specific features of the environment. As humans have lived in such variable environments, many high-load problems cannot be solved by prewiring information into human heads. Our genes cannot predict the kind of world in which we will live. That has been true for at least 200,000 years, probably longer. The spread of variation across space and time is large. In some cases, we may have access to adaptively salient information by being prewired with the much of the information we need (or with partially specified schemes allowing learning to fine-tune such prewired capacities). The physical properties that govern mechanical interaction between stone, bone, and other hard materials are important to our lives as tool-assisted foragers, and we have some evidence that these physical principles are difficult to grasp. Chimps learn to manipulate their environment by trial and error, but they do not automatically generalize from one setup to a causally similar one (see Povinelli et al. 2000, though this interpretation is controversial; see Herrmann et al. 2007). These mechanical principles are common to, and adaptively important in, all human environments, and so humans may well be pre-equipped with information about mechanical causation (Pinker 2007). But this model does not export to most other competences. Even if we confine our attention to humans before the invention of farming and domestication, humans have experienced and adaptively responded to ecological challenges as varied as hot inland deserts (central Australia), extreme seasonal variation (the high Arctic) tropical rain forests (Africa, Central America), shallow tropical seas (the Indonesian archipelago), and large-game specialization (the African savanna). While some principles of biology and naive physics are constant across the ecological challenges those environments pose, the constant features are extremely coarse-grained. Most of what these different peoples need to know will be *specific* to their circumstances.

Moreover, ecology, demography, social structure, and specialization interact. The differences in ecology ramify. These foraging peoples live in different social and psychological worlds, not just different ecological worlds. The problem of novelty cannot be contained to a single domain. Changes in ecology and demography are reflected in changes in specialization, stratification, and investment in high-fidelity cultural learning. These in turn affect the social and psychological judgments an agent must make. For example, the problem of trust changes as we shift from relatively homogeneous and intimate social worlds to those in which differentiation and exchange play a more central role. As social stratification becomes important (and grave goods hint that some forms have deep roots), social and sexual decision making has ever higher stakes, as the differences between winners and losers become more marked. Defection and deception become serious dangers (more on this in sec. 6.3). As group size increases, or as interactions with other groups become more common, interactions with relative strangers grow in importance. For example, the social worlds of the complex foraging societies of the Pacific Northwest, organized around salmon exploitation, differed greatly from those of the Australian Aborigines of the first twenty thousand years of their occupation (Brumm and Moore 2005; Habgood and Franklin 2008; Keen 2006; O'Connell and Allen 2007). The societies of the Pacific Northwest had a highly developed technology and intricate systems of public symbols; they were densely populated, with marked social stratification. The early Aboriginal world had low population densities with small, scattered groups; a limited technology; and few signs of social stratification and public symbol use. Yet many were sexual geron-tocracies with extraordinarily elaborate norms of sexual access (Keen 2006). The problems of social navigation and mind reading in ancient Australia differed from those of the Pacific Northwest.

In sum, then, the organizing theme of the next few chapters is the problem of novelty and the idea that we solve that problem by being able to accumulate and use cognitive capital. The story of hominin evolution is not the story of the evolution of specialized, innately structured modules. Nor is it a story of a key innovation and its consequences. As we will see over the next few chapters, it has often been suggested that the ape–hom-inin divergence is due to a specific adaptive breakthrough in our lineage. Three recent suggestions include fire and cooking, the use of weapons, and cooperative breeding. These were important, but the model I develop here emphasizes positive feedback loops between many aspects of hominin life, for the hominin adaptive complex involves many features that are novel to, or greatly exaggerated in, our lineage. These include technology, vari-ous forms of ecological cooperation, and a transformed and complex social life. I suggest that these hominin specializations began to evolve early, per-haps around the australopithecine–habiline transition, and with change becoming more marked with the origin of the erectines (perhaps around 1.8 million years ago). In many respects, the dates remain conjectural. More important, these hominin novelties coevolved in positive feedback loops, or so I argue. There is no master adaptation whose origin explains the rest.

In chapter 2, I introduce a basic model of accumulation and apply it to the broadest outline of the evolution of hominin social learning in chapter 3. In that chapter, I illustrate the explanatory power of the basic model by using it to explore a range of puzzling phenomena in archaeology, includ-ing Neanderthal extinction. I link information sharing to other forms of cooperation in chapter 4. Chapters 5 and 6 elaborate the basic model by linking it to issues in signaling, trust, and communication. I remarked in section 1.2 that the defection problem is overemphasized in some models of human evolution. But it is important, and chapters 5 and 6 explore the connections between cooperation, communication, and the suppression of cheating. Many forms of late hominin cooperation depend on trust: on an agent's capacity to make credible threats and promises. Credibility, in turn, has been seen as a signaling problem, the problem of ensuring honest communication between the trustworthy in the face of the threats of deception. So these chapters explore the mechanisms that make honest communication possible, and the mechanisms that often make it possible to trust the promises and threats of others. The basic model of social learning built in section 2.3 focuses on the evolution of the capacity to transmit skill and expertise across the generations, so chapter 7 extends that basic model by applying it a broader range: the intergenerational transmission of norms and ideology. The discussion of cooperation and threats to cooperation in chapters 4, 5, and 6 focuses on interactions within local groups or communities. Chapter 8 takes up the issue of relations between groups and the idea that cooperation within human groups depends on cutthroat competition between them. I am skeptical, and I explain why. The chapter ends by identifying important open questions, and with a final overview.

# 2  Accumulating Cognitive Capital

Accumulate, accumulate! That is Moses and the prophets!
—Karl Marx, *Capital*, I, chapter 24

## 2.1  A Lineage Explanation of Social Learning

This book is at heart an attempt to explain the origin and special role of cultural learning in human evolution. That is not because I think the expansion of social learning is the key innovation whose origin drove hominin evolution. Social learning, advances in technical capabilities, changes in prosocial motivations, and cooperation in breeding were all important, and they all interacted. That said, the evolution of accumulating social learning was one central causal factor in the evolution of human uniqueness. Its evolution did not just change hominin phenotypes; it changed the way the hominin lineage evolved. Almost certainly, the earliest hominins, like the living chimp species, resembled their parents largely as a result of the flow of genes across the generations. To a reasonable approximation, with them, inheritance was genetic inheritance.[1] Hominins evolved new inheritance mechanisms, not just new phenotypes. Late hominins resembled their parents in part because genes travel across the generations; in part because the preceding generation engineers the developmental environment of the next generation; and in part because information, mediated by social learning, flows from one generation to the next. Multiple inheritance mechanisms transformed human evolutionary regimes (I return to this idea in sec. 8.1).

Cultural learning plays a core role in hominin life, in part because humans and their ancestors lived in changing worlds. As we saw in sections 1.1 and 1.5, environmental change has been intense, pervasive, and persistent. Since hominin environments were variable over space and time, hominins needed to learn about their environment. In contrast, gorillas do not need to learn that they will grow up in a world of dominant males with harems; that knowledge can be implicit in their default behaviors, because it is a constant of gorilla life. Hominin lives had few such constants. Moreover, often, what hominins needed to learn, they could learn socially. The environment did not change so fast that parental information dated too rapidly to be useful. It is a truism of evolutionary theory that moderate rates of environmental change favor social learning, for while it is sometimes less reliable than individual trial-and-error learning, social learning is much cheaper; others bear the cost of experiment (Boyd and Richerson 1996; Laland 2001). Many aspects of hominin environments do seem to have changed at such moderate rates.

However, social learning has been important, not just because the pace of change has favored exploiting the skills of others; it has been important because hominins have needed to know so much. Change has not just been pervasive. It has also been multidimensional. Hominin social environments have changed in size, degree of individual specialization, life history, patterns of intergroup relations, forms of hierarchy and within-group control, sexual division of labor, and the role of norms and ritual. Some of these changes have probably been slow enough to allow genetic accommodation (for example, changes in life history). But other features of the environment vary from group to group at a time and can presumably change for a group over a few generations. So, for example, the extent to which human reproduction takes place in the context of long-term, monogamous partnerships is quite varied across human cultures. So too is the sexual division of labor and family organization.

In addition, humans have come to depend on many kinds of resources, and that dependence makes many physical and biological dimensions relevant to human fitness, each of which generates learning demands. Habitat patches that differ only in the availability of flint or ocher are equivalent to chimps but were not to our ancestors. We needed to learn where the flint was; chimps did not. Finally, collective foraging requires not just technical and environmental expertise but social skills as well. These interpersonal tools are often acquired socially as agents acquire the coordination and negotiation tools that make social foraging possible. These tools include norms and customs, action-controlling signals, and sensitivity to relevant social information. By two hundred thousand years ago, human children had a lot to learn. The result, as I argued in the last chapter, has been a revolution in the nature and role of social learning in hominin life, and the evolution of a cluster of adaptations that support such learning.[2]

To explain this extraordinary growth in intergenerational social learning, we need to understand the selective advantages that flow from enhanced learning. But, as Brett Calcott (2008a) has shown, we need as well a "lineage explanation" of the evolution of cultural learning: an explanation that specifies a sequence of minimal, incremental changes that takes us from agents with baseline capacities for social learning (perhaps capacities roughly similar to those of living great apes) to agents whose cognitive lives are organized around social learning. I outline a lineage explanation of human social learning in this chapter; much of the rest of the book extends this outline and explores its ramifications. The explanation I outline decomposes the vast difference between modern human capacities and baseline capacities into small increments. But it has five other features: (i) it shows how social learning preceded adaptations for social learning; (ii) it shows how the scope of social learning can expand over time; (iii) it shows how social learning, once expanded, depends on adapted environments, not just adapted minds; (iv) it shows that social learning is often a hybrid of social and trial-and-error learning, and that this is important to the power and reliability of social learning; and (v) it shows how social learning interacts positively with other forms of cooperative interaction. In the rest of this section, I expand on these rather compressed remarks.

*First learning, then adaptations for learning.* The lineage explanation begins by trying to picture the establishment and stabilization of cross-generational cultural learning in agents who are not specifically adapted for such learning. In doing so, I have adapted ideas from West-Eberhard on the importance to evolution of phenotypic plasticity and from the niche construction literature. Thus, in showing that phenotypic plasticity is central to evolutionary change, West-Eberhard (2003) argues that genes are the "followers rather than the leaders in evolution." The idea is that environmental changes often induce new phenotypes by existing mechanisms of developmental plasticity. Genes have norms of reaction: their effects are often sensitive to specific features of the environment in which they are expressed, and so environmental change can expose previously hidden genetic potential (Lewontin 2000). Environmental change thus often brings with it a changed population, and that in turn also changes the selective environment. Existing genes that promote phenotypic accommodation to the new environment—perhaps one promoting resistance to a new toxin to which the population is exposed—will become more frequent; new variants that also promote accommodation will be preserved and amplified. So if the environmental change persists, there is likely to be genetic adjustment to it, initially by selection of existing genetic variation, eventually supplemented by new genes. I argue that the initial establishment of cross-generational learning was probably a special case of a West-Eberhard transition. It begins without special adaptations for cross-generational cultural learning, but these evolve as a consequence of the establishment of cultural learning as a central feature of hominin lives.

*The scope of social learning expands.* One aspect of the transition is thus a shift from agents whose social learning depended on cognitive mechanisms that evolved for other purposes (individual trial-and-error learning) to agents with multiple adaptations for social learning. A second aspect of this transition involves the kind of information that flowed cross-generationally. I suspect that early hominins largely acquired procedural information from their parental generation. They acquired parental skills. Late hominins learn about much more; in particular, they acquire from the parental generation much more declarative information (a special case is the focus of chapter 7). The expanded scope of cross-generational social learning is intimately connected to the establishment of new channels through which information passes. The connection between scope and channel is the main focus of section 6.3. I argue that there is an intrinsic constraint on the order of appearance of channels of cross-generational information flow. Early channels are error tolerant and offer few opportunities for deceptive exploitation. They are safe for agents that lack sophisticated adaptations for social learning.

In particular, as our capacities for social learning evolve, late hominins learn socially some of the tools for social learning itself, most obviously language (but also many theory-of-mind skills). As Andy Clark (2002, 2008) and Dan Dennett (2000) have argued, tools for learning and thinking play a crucial role in human cognitive evolution and human cognitive performance. The invention of numerals, and of systems of numerical notation, enabled humans to think about quantity in ways that were previously impossible (Dehaene 1997; Everett 2005; Frank et al. 2008). We came to be able to represent and reason about large numbers with precision. Material symbols enhance memory, as do various other external prompts (Clark 2008). Many important cognitive capacities are like literacy: they depend on material culture and exist only in environments in which they are supported. Such tools become especially important when environments change at rates that render obsolete the expertise of the parental generation. In such fast-changing domains, our capacity to navigate novelty depends heavily on cognitive tools. We use epistemic technology to organize and store information in accessible and user-friendly forms and to simplify our information environment (color-coding wires, for example); we use technology to lower the costs of exploration (using a stick rather than a finger to probe a hole or remove an obstruction). We use technology as cognitive prosthetics, for example, by using tally marks on a wall to expand memory. The model of social learning and its evolution that I develop in section 2.3 gives a smooth account of the increasing importance of cognitive tools. Initially artifacts and actions have informational effects as side consequences of more directly utilitarian functions. As Dennett has noted, a wheeled cart carries the idea of a cart along with its physical load. These are stepping-stones to specialist epistemic action (as in marking trails) and epistemic artifacts, as with calendars, and tally sticks.

*Adapted social environments.* Social learning has been important in the hominin lineage for a long time, perhaps as early as Oldowan technology (around 2.3 million years ago, perhaps even earlier, at 3.4 million years ago, if a recent report proves correct [McPherron et al. 2010]). As a consequence of its importance, individual cognitive adaptations have evolved in the lineage. But the social environment, not just individual minds, has become increasingly organized to support the flow of information across the generations. Learning is supported by the provision of resources to children, giving them the time and space to explore and to absorb and practice skills. They are protected, ensuring that their trial-and-error exploration is not catastrophically dangerous. In many cultures, children's toys and games teach crucial skills. More generally, children's social learning is enhanced by tolerating their inquisitiveness about adult activities, by allowing them opportunities to explore adult material culture, and by providing children with advice and explicit instruction. All these practices vary significantly from culture to culture. But cultures provide children with informational as well as material support.

*Social learning is (often) hybrid learning.* Sometimes social learning is direct. I want to know how to solve a problem with my computer, and the help-desk adviser tells me where to find the crucial command in the menu; I want to know how to run my iPod in "shuffle" mode, and my daughter shows me the right command. Most of the recent, expanding experimental literature focuses on such cases of pure instruction, or pure demonstration, for example, in testing the reliability of transmission chains under various conditions (for a review, see Mesoudi and Whiten 2008). Many studies of social learning in children focus on the fidelity with which information flows from one child to another in diffusion chains (Flynn 2008). But the most consequential cases of social learning in humans have not depended on pure demonstration or instruction (except, perhaps, in the last few thousand years). Rather, most social learning is hybrid learning: agents acquire skills through socially guided trial and error and socially guided practice. Children do get advice, instruction, and other informational head starts from others, but they get this support while engaged in exploratory learning in their environment.

*The cooperation syndrome.* Sharing information is a special case of cooperation, so it is important to show how information sharing coevolves with, and interpenetrates, other forms of cooperation. The connections between informational and ecological cooperation are central to chapters 4 and 5. Chapter 4 responds to a skeptical challenge. The model of human evolution developed here presupposes that cooperative foraging has long played a central role in human life; it presupposes that cooperative foraging was central to hominins becoming human. An alternative model identifies grandmother–daughter reproductive cooperation as the key innovation of human cooperation and is skeptical about the idea that hominin life history patterns evolved to make room for large intergenerational transfers of cognitive capital. I doubt that reproductive cooperation was a key innovation driving hominin cooperation. But I do think it was part of the mix, and in chapter 4 I sketch an integrated model in which informational, reproductive, and ecological cooperation coevolve under positive feedback. I take up this idea of feedback, and its role in the expansion of social learning, in the next section.

## 2.2   Feedback Loops

I suggested in section 1.5 that it was more productive to think of hominin evolution as being driven by positive feedback loops than by a key innovation. Coevolution, not a magic moment when a special light turns on inside, explains why we are so very different from the other great apes in our behavior, our cognition, and our social organization. In this section, I identify two important loops: one between expertise, social learning, and life history; the other between expertise, individual adaptations for social learning, and organized learning environments. The first loop is acknowledged but controversial; the other, largely ignored. These loops help explain the coevolutionary interactions between informational and other forms of cooperation, the expanding bandwidth of social learning as it comes to play a role in more and more competences, and the increasing fidelity of cultural learning as particular skills come to be relearned more accurately as a result of increasing social input. Fidelity is of particular significance, because there is a widely held (though contested) view that the character of culture changes once a fidelity threshold is reached. Cultures become accumulative as fidelity allows existing competences to be improved incrementally. That thought contains an important insight, though as we shall see, the relationship between fidelity and accumulation is complex. In the next section, I use these ideas to develop a basic model of the evolution of human social learning; in this section, I focus on the feedback loops themselves.

I begin with the connections between informational cooperation and life history. Social learning in many species consists in animals using one another's behavior as cues:[3] one bird flies down to a possible food source, and others follow it. In contrast, we absorb much long-shelf-life information from the previous generation, often as a result of intentional information generation (Danchin and Luc-Alain 2004; Laland and Hoppitt 2003; Reader and Laland 2002). This high-volume, high-fidelity, cross-generational information flow depends on the extension of human childhood, the invention of adolescence, and the concomitant extension of human life spans. Those changes in human life history, in turn, depend on ecological innovation. Selection for social foraging selected for information pooling at and across generations, and these in turn supported increased specialization and division of labor. Children are expensive. Our foraging ancestors could afford them only because they spent their own childhoods acquiring skills that enabled them, in turn, to support children. Cross-generational information flow both depends on and helps sustain the cross-generational flow of physical resources. We live twenty years longer than great apes, and the extension of human life span and the investment in a long childhood depended on controlling extrinsic causes of mortality: reducing the risks of predation and accident, and reducing the risk of starvation when ill, injured, or unlucky (Hill and Kaplan 1999). The extension of lifespan depended as well on the parental generation being able to support the next generation until they are themselves ecologically competent. Human social learning depends on the fact that adults (and especially adult males) generate far more resources than they consume themselves. For human populations to be viable, the adults of generation N+1 must survive long enough to pump resources to generation N+2 equivalent to those they received from generation N. Socially learned information makes adult foraging safe and profitable; because adult foraging is safe and profitable, adults can afford children who remain partially dependent for many years.[4]

So human life history characteristics coevolve with technological competence and cultural learning. The technological and informational bases of cooperative technological foraging typically require deep educations. In many cases, foragers do not peak in their resource acquisition powers until they are about thirty; they do not begin to produce more than they consume until about eighteen. The resource debt that individuals acquire as children and adolescents is not paid off until around fifty.[5] Extensive cultural transmission supports the development of technological foraging skills that generate extremely rich returns; the life history that makes such cultural transmission possible is paid for by those same skills (Robson and Kaplan 2003; Kaplan et al. 2005; Gurven and Hill 2009; Kaplan, Hooper, and Gurven 2009).

There is a second important feedback loop: an interaction between environments organized for social learning and individual cognitive adaptation. Cross-generational cultural learning need not depend on specific adaptations for social learning. As Eryton Avital and Eva Jablonka have pointed out, a fortuitous individual innovation that gives access to a new resource can reorganize the lifeway of the innovating animal and its associates as they exploit that resource. If young animals associate with their parents, this new way of life automatically alters the environment that the juveniles explore. As a consequence, their trial-and-error learning will be rich with opportunities to learn about the new resource. So a lucky accident at generation N becomes a regular event at N+1. Among other examples, Avital and Jablonka show that the celebrated case of potato washing by Japanese macaques probably spread by this mechanism (Avital and Jablonka 2000; Jablonka and Lamb 2005).

The potential power of this mechanism is surprisingly easy to overlook. So, for example, Tennie, Call, and Tomasello (2009) miss it in analyzing the difference between accumulating human and impoverished chimp cultures. They recognize that adult activities can seed juvenile learning, noting that chimp "nut crackers and termite fishers leave their tools and detritus behind, and in the right places, which make the learning of their offspring and others much easier" (2406), and so "once an individual has invented one of these rare types of behaviour, its activities or resulting products make it considerably easier for others to repeat the behaviour, and this they do basically on their own by mostly re-inventing it" (2407). Even so, these authors regard individual chimps as basically on their own. Since chimps lack cognitive adaptations for high-fidelity cultural learning, their social worlds are constrained. Chimps can innovate, but only to capacities that are a short learning step away from chimp defaults, so that chimps live within a "zone of latent solutions." In chimp communities we see only "things that individual chimps could invent on their own fairly readily." And so we do not see much diversity across chimp communities; initial chance differences in innovation do not accumulate, leading to increasing divergence over time, as we find in human cultures (Tennie, Call, and Tomasello 2009).

However, if seeding by the previous generation is important, then chimps (and early hominins) do not have to invent these new skills or artifacts "on their own." Learning mediated by the environmental effects of one's social group is not solitary learning, even if the internal cognitive mechanisms deployed are the same as those used in exploration learning. This is not just a semantic point: Tennie, Call, and Tomasello use their analysis to predict a fixed limit on ape cultures, which "are restricted by the upper boundaries of their species ZLS [zone of latent solutions]" (2407). But even without special mechanisms of cultural learning, seeding by the previous generation can be *chained* to produce a sequence of adaptive changes that transform a lineage's lifeway. Imagine the gradual invasion of a new habitat. Initially, suppose adults learn to exploit seasonally abundant resources in an adjacent ecotype. They occupy that ecotype for longer periods as a result of this entry point, and hence exploration learning gives them a chance to discover and exploit further resources, to learn about and avoid new dangers. This could well be an incremental process in which each innovation, by increasing the time spent in the zone, makes further innovation more likely while making the acquisition of new competences by the next generation a matter of routine. Think, for example, of wetlands. They sometimes offer rich and obvious pickings (for example, when water levels drop). These windfall temptations can then generate opportunities to explore and innovate. It may be that the incremental improvement of an existing skill or competence requires dedicated mechanisms of social learning. But accumulation by serial addition—increasing the bandwidth rather than the fidelity of cultural learning—need not.

This is not an idle speculation; I think this process was probably important in the establishment of material culture in the hominin lineage. The first traditions of using stone technologies may well have been established by such an epistemic side effect of innovation. Even the simplest flake-and-core technology is not cognitively trivial. Stones were probably first used simply as hammers, to break into nuts and bones to extract their contents. Great apes can use hard objects as hammers to break containers open, and so early hominins were probably capable of using stones as hammers. Making the simplest cutting tool is a significant advance on using stones as hammers; it requires greater skill to produce a sharp flake, for the toolmaker must control the strike angle and impact zone. So the flake maker needs both extra information and more physical control. To shape requires more precision and control of force than to break (de Beaume 2004). Chimps can be taught to make sharp flakes, but comparisons between the debris produced by chimps, skilled flint workers, and those from archaeological sites suggest that Oldowan toolmakers were more skilled than chimps, using less brute force in their approach to their tasks (Whiten, Schick, and Toth 2009). So Oldowan technology is a skilled technology, despite its rough-and-ready appearance. Still, agents who regularly explore the mechanical properties of stone and experiment with the effects of the angle and power of impact will, if they persist, find ways of making flakes. Young habilines would have had such opportunities. If using cores and flakes gave adults access to, say, carcasses and marrow, their children would have many chances to explore and play. The archaeological record shows that making and using

Oldowan tools was a regular activity rather than an occasional one; in some sites, cores show signs of heavy and repeated use (Whiten, Schick, and Toth 2009, 2).

So information-rich traditions might become quite important without any change in individual cognitive equipment. The social environment changes as a result of innovation or environmental change, and as a consequence, parental capacities are reliably regained in the next generation. But they are regained through adults' shaping juvenile learning environments as a by-product of the adults' own ecological activities. They are not regained as a result of adaptations for social learning, or teaching. *No feedback loop yet.* But clearly, once information-rich traditions are established, the selective environment changes. The initial shift to a stone-tool-based lifestyle may well have depended on preexisting mechanisms of adaptive plasticity, preexisting potentials for manual dexterity, and preexisting foraging patterns. But once established, the new lifestyle will select for genetic variants that enable these new skills to be acquired with high reliability and low cost (it is easy to lose eyes and fingers while flint knapping). There will be selection in favor of mutations that increase the reliability and accuracy of learning from the parental generation (unless these mutations come with other, unaffordable costs). Such mutations can adapt morphology as well as mind to the new technology. As Ambrose (2001) notes, our hands, wrists, and arms are better suited than those of chimps for flint knapping. Likewise, if traditions are transmitted vertically from parent to offspring, mutations that alter parental behavior in ways that increase transmission reliability will be favored.

This process might begin with simple motivational changes. Parents might become more tolerant of juveniles in close proximity, and tolerant of them experimenting with adult possessions—tools not in use, fragments of newly extracted resources. Juveniles might become more focused on adult activity. There is reason to believe that New Caledonian crows, celebrated for their use of pandanus tools for extracting insects, have adaptations like these. Adults are tolerant, and juveniles are curious about adults and their gear (Russell Gray, personal communication; see also Holzhaider et al. 2010a,b). Even modest fine-tuning of individual cognitive phenotypes to improve their capacity to extract information from the previous generation will increase the potential bandwidth of cross-generational learning. As learning becomes quicker, more reliable, and more extensive, a generation can accumulate more cognitive capital, because they take better advantage of what their predecessors knew. As they know and can learn more, they can spread into new habitats and exploit a still broader range of resources, thus making still more information relevant to their lives.

Of course, as noted earlier, it is always possible that genetic changes that increase the efficiency of cultural learning have countervailing costs. It has sometimes been thought that Baldwin effects are rare, because bringing a learned response under genetic control has the hidden costs of reducing the overall adaptive plasticity of the agent (Weber and Depew 2003). I do not think that a similar trade-off occurs between individual and social learning. Indeed, a positive feedback loop may even connect the two forms of learning. If social learning enables a young hominin to acquire his or her basic skill set reliably and efficiently and while still a juvenile (hence still with the opportunity and motivation to play and explore), then such an agent will have more opportunity to explore and experiment, while already equipped with skills that make an adult competent. Social learning could provide, rapidly and efficiently, a platform that individual learning might upgrade and extend. Moreover, in part social and individual learning depend on shared mechanisms: memory, the control of attention, an ability to inhibit impulse, and the ability to monitor the results of one's own actions. In such cases, genetic variations that enhance social learning by improving these mechanisms will enhance individual learning too.

As cognitive capital at a generation becomes increasingly important, there is increased selection for cognitive capacities that enable juveniles to pump information from their seniors. At some stage deep in our history, hominin lives became utterly dependent on the reliable preservation of these rich bodies of information. In our lineage, traditions of expertise became more information rich and more numerous, and so we have evolved specific and complex adaptations for social learning. More on these in chapter 6. The purpose of this chapter is to introduce a basic mechanism of cross-generational transmission, one that can begin without initial adaptations for that transmission but one that has been elaborated in ways that increase its scope, fidelity, and bandwidth.

## 2.3   The Apprentice Learning Model

Apprentice learning offers a helpful conceptual model of the synergy between organized learning environments and individual cognitive adaptations. Apprentice learning is a powerful mode of social learning, making possible the reliable acquisition of complex and difficult skills, as craft apprenticeship learning shows. Skill acquisition in forager societies is, I think, often similar to apprentice learning, and that explains the possibility of high-volume, high-fidelity social learning in these cultures. Apprentice learning is hybrid learning: combining information from the social world with information from the physical-biological environment. It is learning by doing. But it is learning by doing in an environment seeded with informational resources. These include raw materials, in both their raw and processed forms. In addition, full and partial templates of the final product are available to guide action. So too are tools. Moreover, there are many opportunities to learn by observing highly skilled practitioners. Advice is often available from both experts and peers, for learning is often social and collaborative. Indeed, the learning trajectory of an apprentice is often at least partially organized by experts. The expert organize the trial-and-error learning of the less expert by a combination of (i) task decomposition and (ii) ordering skill acquisition, so that each step prepares the next. Often, even in traditional cultures, apprentices must show commitment to learning and pay for information by performing unskilled and semi-skilled labor (Stout 2002). So ordering skill acquisition is typically in the interests of both expert and apprentice. It is in the interests of the expert to get as much work as possible out of the apprentice, so it is in expert interest to assign tasks up to, but not beyond, their skill ceiling; expert interest thus structures apprentice practice adaptively. Thus apprentice learning depends on individual cognitive adaptations for social learning but depends as well on these adaptively structured learning environments. I argue that this mode of social learning has deep roots in hominin history. In my view, craft expertise—the kind of skill sets that forager lives depend on—is fine-tuned at a generation and reliably transmitted across generations by this mode of organized human learning environments. If so, apprentice learning is a good general model of the acquisition and exercise of the crucial cognitive competences that mediate our adaptive responses to a complex and changing world.

The apprentice learning model has four important virtues. First, it identifies a form of learning that can be assembled incrementally. The reliable transmission of skill can begin as a side effect of adult activity, without adult teaching and without adaptations for social learning in the young. Once established, it then brings with it selection for cognitive and social changes that increase the reliability or reduce the cost of learning. Rudimentary but reliable skill transmission, however, does not presuppose the presence of such adaptations (Avital and Jablonka 2000). Second, apprentice learning is known to support high-fidelity, high-bandwidth knowledge flow. Until recently, much technical competence in industrial society depended on apprentice learning. Virtually all technical competence in preindustrial societies depended on it. Third, the model fits ethnographic data quite well. Formal educational institutions and explicit teaching are not prominent parts of traditional society. But many forager societies organize and enhance children's participation in economic activity, and this approach supports the transmission of traditional craft skills (Bock 2005; MacDonald 2007). Finally, the model can be shown to illuminate the archaeological record, or so I argue in the next chapter. Let me now say a little in support of the idea that this basic model has broad application beyond formal apprenticeship in highly skilled craft guilds, before I turn to incremental construction and ethnographic plausibility.

*Apprentice learning does not require explicit instruction or formalized institutions.* In considering the range of cases we can understand through the apprentice learning schema, it is important to realize that experts can structure the learning environment of the inexpert without much explicit teaching. As well as seeing expert practitioners in action and helping them, children often have a chance to listen to experts talk to one another and to more skilled apprentices about their expertise. Listening in helps in acquiring local lore as well as local practices. They learn from one another as well. The general picture is that much skill learning in forager society is accomplished by trial and error, but supervised and organized trial and error. Moreover, it is trial and error in an environment seeded with props and other cognitive tools. The specialist vocabulary to which children are exposed marks salient distinctions. Tools and artifacts—finished, half finished, and broken—are available as sources of inspiration and comparison. In short, while the role of explicit teaching in traditional societies is often quite limited, adults can and do structure and engineer the learning environment, even without explicit teaching.

That said, in many cases, teaching may well be important. Lyn Wadley (2010) shows that Middle Stone Age composite technology depended on glues that were difficult to make and use, and she suggests that these skills could be transmitted only by explicit teaching using language. Likewise

Tehrani and Riede (2008) point out that the manual skills required for many traditional craft skills are highly intricate, so much so that they would be difficult to master by imitation learning. Yet often such skills are transmitted so reliably that characteristic products reappear recognizably for many generations. Jan Apel (2008), for example, details intricately made Neolithic stone daggers from Scandinavia made to a design that was transmitted for at least twenty-four generations. Tehrani and Riede argue that such cases show the historical depth of active pedagogy. I agree, but teaching often consists in enriching the learning environment as well as direct instruction.

Phenomenologically, specific expert skills are somewhat akin to the modules of the standard model. Think, for example, of literacy or of the precise quantitative reasoning that our mastery of the integers and positional notation gives us. These are in some ways our equivalent of the skills in forager skill sets. We see the appropriate physical patterns as words rapidly, automatically, and without conscious effort; we can do so while being engaged in other tasks. Once the skill is fully acquired and installed, reading is no longer difficult; it no longer demands attention. But while phenomenologically, literacy and simple numeracy are akin to modular capacities, they are not so developmentally. Literacy is not triggered by experience: it is acquired relatively slowly, with considerable effort and variation. It is acquired reliably only in informationally enriched environments. Literacy is not a prewired adaptation. Indeed, literacy illustrates our adaptive response to novelty. In contemporary mass societies, it is crucial. But the features of the world to which literacy is a response—frozen language and long-distance decontextualized communication—are novel. Yet most individuals in First World societies become functionally literate: they are able to act adaptively in a world in which much language is not speech.

*The incremental construction of apprentice learning.* One advantage of the model is that apprentice learning can evolve incrementally, from a baseline in which juveniles have quite minimal individual adaptations for social learning, and where adults have no cognitive adaptations for teaching, but where the social environment favors skill transmission as a by-product of adult economic activity. Thus a full apprentice model of expertise transmission does not apply to early hominins. Unlike their distant descendants, they are not equipped with a rich set of individual cognitive adaptations for social learning. Nor in all probability did they develop in environments adapted for social transmission. Nonetheless, on the assumption that young hominins spent a good deal of time with their elders, and on the assumption that those elders were tolerant of children's presence, curiosity, and experimentation, the basic hybrid learning model applies to early hominins. They too learned by doing, in environments that advantageously shaped individual trial-and-error learning. For according to these assumptions, the learning environment of young hominins is structured advantageously by adults through the exercise of the adults' own expertise. Tools, partially completed tools, and raw materials were readily available as objects of play, experiment, and exploration. Even if the young hominins of 2.5 million years ago were not highly competent observational learners, emulation was surely within their capabilities. They could see, for example, that rock was a resource and that sharp flakes were its product, even if they could not learn just by seeing just how the flakes were struck from the core. So their exploration would be focused advantageously by emulation. The same is likely to be true of any resource whose exploitation depended on skill that played a central role in adult life. If tubers and other underground storage organs were indeed important dry-season resources for hominins 1.5 million years ago (an idea we will meet in chapter 4), young hominins would have many opportunities to experiment with the skills needed to find and process them. Hybrid learning of this kind might not have been high fidelity, and its bandwidth was surely limited. But it would be reliable: some version of a few crucial skills would be acquired by almost everyone in the next generation.

*Ethnographic plausibility.* A further advantage of the model is its ethnographic plausibility. Our equivalent of craft skills, as I noted earlier, are intellectual skills like literacy and numeracy. These skills too are acquired by supervised and structured practice, though in ways that depend on formal educational institutions. In this respect, they contrast with forager skill sets. But while foragers do not depend on formal educational institutions, they engineer their children's learning environment (more on this in chapter 4). So, for example, forager children are provided with toys (for example, miniature bows) and are encouraged in games that train them in crucial skills. These games often rehearse key physical skills. In their study of Congo Basin foragers, Barry Hewlett and his colleagues note that parents provide children with effective but miniaturized tools (axes, digging sticks, baskets and the like) and encouraged and advised on their use (Hewlett et al. 2011, 1174–1175). Likewise, Australian Aboriginal children are provided with hunting toys (Haagen 1994) and are sometimes taught how to make the tracks they must follow (Morrison 1981, 166–176). Forager villages often keep a large range of semiwild pets, and this gives children the opportunity to learn the animals' calls, behaviors, tracks, and scats.

In many forager societies, children, especially somewhat older children, contribute to the family economy. But to allow them to do so, they are provided with equipment appropriate to their size, strength, skill level, and local ecology: fishing lines or spears, nets, baskets, and the like. They learn by doing, but what they do is engineered by adult experts via their equipment supply. Children are taken on adult foraging expeditions, and these are sometimes modified to make the trips safer or more educational for the children (MacDonald 2007), and on these trips, these children are exposed to an enormous amount of hunting lore just through being part of adult conversational circles (Marlowe 2005, 2007).[6] Indeed, in some cultures hunting skill is passed on through something like explicit apprenticeship (Bahn 2007). Children often begin to learn craft skills by first helping their adult relatives, combining practice with observation: again, learning by doing, but with skilled adults organizing the sequence with which skills are acquired (see Bock 2005; for a more skeptical view of the informational demands of hunting, see Bliege Bird and Bird 2002).

There is ongoing controversy about the role of explicit teaching in traditional society, for there has been an influential line of thought suggesting that it is essentially absent from such societies. But Csibra and Gergely (2011), in their recent response to this literature, argue persuasively that it often depends on a highly formalized, institutional conception of teaching, not counting, for example, informal advice and encouragement as teaching. This message is reinforced by Hewlett's review of the Congo Basin studies. Those studies rely on both detailed observation and interview, and they show that explicit teaching plays an important role in this foraging culture. So, for example, in one set of examples, Hewlett describes interviews in which Aka women recalled how they learned to distinguish edible from inedible yams and mushrooms. When they were young, their mothers laid out inedible and edible forms in front of them, and explained how to distinguish the safe from the dangerous varieties (Hewlett et al. 2011, 1175). So foragers do deliberately teach their children. But even in the absence of such explicit teaching, children are supported with informational resources, and learn in an enriched environment. Whether or not explicit teaching plays a prominent role, high-fidelity, high-bandwidth social learning depends both on an organized and adapted learning environment and on specific cognitive adaptations.[7]

The ethnographic plausibility of the model is supported by the literature of anthropology. For it details many examples of craft skill transmission (particularly weaving traditions) in apprentice-style situations (Tehrani and Riede 2008, 321–322). Jean Lave (1996) discusses two examples in some detail: apprentice tailors in Liberia and the study of Islamic law in nineteenth-century Cairo. These examples are important because they document the flexibility of apprentice learning and teaching; this mode of learning supports the acquisition of much more than manual skill. Liberian apprentice tailors learn about the social and economic organization of a tailor's life, not just how to make trousers. Islamic law is not a manual skill, but it is not just a textual skill either. The student learns about the social and institutional organization of Islamic courts, not just about the texts, by being immersed in those institutions.

Dietrich Stout's (2002) study of stone adze making in Irian Jaya is one particularly impressive example from the ethnographic literature. The social and informational organization of adze making, as he pictures it, is strikingly akin to a medieval craft guild. The apprenticeship system is quite formal. There is a master adze maker who has at least formal authority over the distribution of raw material to adze makers. Apprentices have to be accepted by a recognized master, and while apprentices are typically close relatives of their master, that is not sufficient. They have to show commitment by doing menial but useful chores until accepted. The apprenticeship system is quite long, typically taking about five years until an apprentice is accepted as an acknowledged master. While an individual is still an apprentice, his work is owned by his master; any profits from the exchange of apprentice-made adzes go the master. These social factors depend on the existence of a real information market: there are definite and important differences between adzes made by the highly skilled and those made by apprentices. Apprentice-made adzes are smaller, because controlling appropriate proportions becomes increasingly difficult as the size of the stone "blank" increases. While teaching is not formalized, stone for the adzes is selected communally, with communal initial working on-site. Then, when the blanks are transported to the village for final working up and polishing, they are shaped communally, with a good deal of public advice and aid from the more skilled to the less skilled (Stout 2002, 702). Such skill transmission is aided by a complex technical vocabulary of stone working that includes specialist terms for types of stone, hammerstone shapes, impact zones for knapping, and the type and angle of strikes. This technical vocabulary allows adze makers to represent their own skills; their knowledge is partially declarative, not just procedural. The parallels with the formal, institutionalized systems of apprentice guilds could hardly be clearer.

It is obviously more difficult to reconstruct the social organization of teaching and learning in extinct cultures. Even so, Stout takes his example to have broad relevance. Somewhat before 1.5 million years ago, the sharp flakes and cores of Oldowan technology were supplemented by a more impressive technology, Acheulian "hand axes." These were more symmetrical and regular in form (though the older ones were less so). In more recent work, Stout suggests that there are important, cumulative improvements in technique through the Acheulian, even though this is not always obvious from the artifacts themselves (Stout 2011) Stout argues that a wide range of late Acheulian and post-Acheulian stone technologies require stone blanks to be extensively shaped using skills that are difficult to acquire. Acheulian toolmaking probably did not depend on the cross-generational transmission of high-fidelity information about the tool itself: nothing like apprentice learning is needed to explain regularity in shape. As McNabb and his coworkers note in their detailed case study of one specific site, we in fact have little evidence of standardization, of group norms governing hand axe design. Plenty of variability is evident in both shape and degree of symmetry (McNabb, Binyon, and Hazelwood 2004), though the extent of regularity in tool shape seems to vary over space and time (McNabb 2005; Lycett and Gowlett 2008). But high-fidelity social learning may well have been important to the transmission of manufacturing techniques. These technologies depend on the careful assessment and selection of raw materials and the removal of large, thin flakes to produce choppers and cutting tools that are not heavy, thick, and cumbersome but have a large cutting edge. In particular, it is difficult to produce stone artifacts in which one dimension is elongated in proportion to another. Thus blade technologies are skill intensive because the artifact is long but thin.

    In brief, Stout suggests that when the skill itself is complex and difficult, we can reasonably infer that it is acquired by an apprenticeship-like system of skill transmission (for a similar argument, see Tehrani and Riede 2008). While that line of thought is plausible, clearly it is inconclusive. Douglas Bamforth and Nyree Finlay (2008) attempt to strengthen the argument by developing criteria for identifying highly skilled stonework and distinguishing it from less skilled work that is likely to be the result of novice practice.[8] In favorable cases, these methods will reveal the presence of high-fidelity, high-volume social learning in former social worlds. They document the importance of social learning. But by themselves they do not show how social learning is organized, nor do they reveal the cognitive capacities it depended on. Fortunately, in a few rare cases, we have more direct archaeological evidence of the social, public organization of production and of the production of nonfunctional "practice" artifacts. Linda Grimm (2000) reconstructs in detail a particular horizon of one site. In doing so, she reconstructs the full manufacturing history of a number of artifacts and the specific locations of their knapping. She suggests that some cores show evidence of being largely made by inexpert knappers, but from blanks provided by experts, and with occasional expert intervention. Her idea is that the inexpert use excessive force and do not get the angles right on impact, so there are identifiable differences between the flakes struck off by experts and those struck off by apprentices. If she is right, she has provided an archaeological exemplar of highly scaffolded learning-by-doing. Like Stout, she suggests that apprentices show commitment and buy their way in by acting as flunkies for skilled masters of technique.[9]

Flint knapping, like most other complex material technologies, requires infrastructure support: preparing workplaces; assembling raw materials and associated products; disposing of waste by-products that would otherwise clutter the workspace; completing a production sequence once an artifact is shaped so that it can be finished through simple, routine actions. So, for example, the adzes made by Irian Jaya masters require hours of tedious polishing to complete once they are essentially shaped. There is grunt work to be done, so apprentices can earn their instruction by providing useful low-skill support. Lyn Wadley's (2010) fascinating experimental reconstruction of Middle Stone Age adhesives supports this general picture. She shows the sophistication of this composite technology. Her glues are based on acacia gum, but the gum needs to be carefully heated and mixed with ground ocher, and once the point(s) are attached to the shaft, the adhesive must be allowed to dry slowly. If the glue is not made or dried properly, the glues variously shrink away from the point, crack, or become brittle. Two messages come through. First, this technology is not easy to discover or to reverse-engineer from the finished product. It is a multistage process with irreversible transformations and no obvious perceptual cues to help the artisan. While acacia gum (which is sticky in its raw state) might be an obvious starting point for glue production, the additives, the heat, and the drying regime are not. Second, as with Stout's adzes, there is real grunt work to be done. It takes about an hour to grind the necessary ocher to attach two points (Wadley 2010). Work for knowledge is a good deal for both sides.

In the next chapter, I provide further support for the ethnographic and archaeological power of this model by showing how richly it illuminates a number of long-standing controversies within archaeology and paleoanthropology. The bottom line, though, is that any good model of the evolution and architecture of the human mind must be built around two key phenomena. The first is that despite the usual examples of birth control and fatty foods, we are not in general incompetent in the face of novelty. Each of us has survived in a world unimaginably different from that of a Pleistocene forager. We need to explain not the existence of adaptive lag but the fact that it is a relatively minor problem. The second is that we survive novelty because we can accumulate and wield cognitive capital. We do not do so perfectly, but we do so well enough to be both numerous and ubiquitous.

# 3 Adapted Individuals, Adapted Environments

## 3.1 Behavioral Modernity

In the last chapter, I outlined an initial model of social learning based on the idea that the upstream generation structures the learning environment of the downstream generation, so that trial-and-error learning combined with observational learning and (sometimes) explicit instruction results in the reliable reacquisition of expertise. According to this model, high-fidelity, high-bandwidth social learning depends both on adapted environments and on adapted minds. The purpose of this chapter is to illustrate the explanatory power of this model, choosing two important paleoanthropological puzzles. One is the puzzle of Neanderthal extinction. The other is the long lag time (perhaps 150,000 years) between the origins of our species and the appearance of recognizable human cultures, cultures that fall within rather than without the variation known from the ethnographic record. It took a hundred thousand years or more for *sapiens* to become "behaviorally modern." I argue (beginning with behavioral modernity) that these puzzles are generated by focusing too exclusively on individuals and individual adaptation. To understand our history and the history of our sister species, it is necessary to focus on the interaction between individual adaptation and social environment, and, especially, the interaction between our individual adaptations and an engineered informational environment. In effect, I argue that the two puzzles illustrate a common dynamic: the importance of thresholds and individual–group interactions. Indeed, I argue in section 3.5 that Neanderthal extinction is a mirror image of behavioral modernity.

The paleobiological literature shows a significant and growing interest in the so-called problem of behavioral modernity (recently reviewed in Nowell

2010a and d'Errico and Stringer 2011). The puzzle arises from an apparent mismatch between the biological origins of our species and the origins and establishment of characteristically human behavioral patterns. Anatomically modern humans appeared on the scene roughly 200,000 years ago, and genetic evidence confirms that our lineage emerged as a distinct, independent lineage around that date (Finlayson 2005; Klein 2009; Tattersall 2009). Yet these First Sapiens behaved (it seems) unlike contemporary humans. Their material technology was much simpler; their foraging breadth was narrower; their social and cultural organization was more rudimentary.[1]

While we have reached broad consensus that early *sapiens* contrast importantly with their later descendants, it is not certain that this consensus is on the money. The problem of behavioral modernity may be a mirage. The supposed qualitative difference between the *sapiens* societies of 200,000 years ago and those of 50,000 years ago might be an illusion generated by an imperfect record of the cultural and cognitive life of the earliest *sapiens*. Peter Hiscock and Sue O'Connor point out that as we look deeper into the past, rare technologies are less well preserved. The record is not just imperfect; it is biased. It shows us less of the most ancient cultures (Hiscock 2008; Hiscock and O'Connor 2006). Moreover, even if the First Sapiens' fundamental cognitive capacities and social organization were the same as those of later *sapiens*, a small, geographically restricted set of populations are bound to have a less varied material technology than larger, more widespread populations. We may be seeing preservation and range size effects rather than real differences. Even so, the apparent differences are so large and long-lasting that researchers generally agree that the cultures of the First Sapiens and Moderns differ qualitatively. I shall work within that consensus.

The nature of that qualitative difference, though, is a matter of great dispute. Until recently it seemed as if the transition from First Sapiens to Moderns was abrupt and coordinated.[2] Somewhere between 60,000 and 40,000 years ago, *sapiens* became human. Technology exploded in regional variation, size of individual toolkits, range of materials used, and complexity of individual tools. At the same time, the economic base of human life became broader. A wider range of animal species was taken. Grains were gathered and ground to flour. Marine resources were added to the human menu, and long-distance trade networks were established. These economic and technological changes were coupled with changes in how humans conceived of themselves in their world. Decoration, ornamentation, and (a little later) musical instruments, cave art, and figurines appear. Researchers first noted these changes occurring in Europe about 40,000 years ago and dubbed them the "Upper Paleolithic Revolution." In Africa, the same sociotechnical complex is known as the Late Stone Age.

While some still argue that the European archaeological record shows a rapid and profound shift (Mellars 2005), researchers now widely accept that many of the traits supposedly definitive of the Upper Paleolithic Revolution appeared earlier (though perhaps spasmodically and in a more rudimentary form) in Africa, perhaps beginning around 150,000 years ago and more regularly between 100,000 and 50,000 years ago. The abrupt European change may well be a sign of migrants moving in rather than abrupt and pervasive innovation. So McBrearty (2007), d'Errico (2011), and Zilhão (2007) all note instances of early use of ornaments and ocher, complex technology (using complex stone technologies and new materials), a widening resource base, and processed plant foods. A good recent review of this material is Conard 2006. The same themes come through repeatedly. Features of material culture and foraging capabilities that were once thought to be diagnostic of the Upper Paleolithic or Late Stone Age turn out to have anticipations in the Middle Stone Age (i.e., roughly 285,000 to 50,000 years ago). So we have Middle Stone Age examples of stone blades, hafted tips, and even standardized tool shapes. In short, we see many early anticipations of, and a slow buildup to, the Late Stone Age or Upper Paleolithic Revolution, if a revolution it was. The material base of technology spread in the Middle Stone Age, though mostly in the second half of that period.[3]

Perhaps the most influential paper driving this change from a pulsed to a gradualist model was Sally McBrearty and Andrew Brooks's "The Revolution That Wasn't." They took behavioral modernity to consist in a cluster of capacities: the capacity to innovate, to plan individually and coordinate with others, to think abstractly. As well, behaviorally modern humans use public, physical symbols in their social interactions with one another. McBrearty and Brooks argued that behaviorally modern human social worlds rest on these capacities, which leave archaeological traces. But these traces appear gradually in the record, at separate times and places. The buildup to behavioral modernity is slow, patchy, and hesitant. It is not pulsed.

A similar message comes from other sources. Just as we see earlier anticipations of the material culture that is distinctive of the Upper Paleolithic, so too we can identify earlier anticipations of Upper Paleolithic economy and social life. Middle Stone Age hominins were active and successful meat eaters. But they also used some plant resources. We have recent evidence that late Middle Stone Age peoples (peoples who lived a bit less than 100,000 years ago) used coastal resources in southern Africa, though ones easily gathered from tidal pools (Marean et al. 2007). So their resource base demonstrates some spread, too. The same report details a similarly early use of ocher, allegedly evidence of early symbolic behavior. The hematite chosen at Pinnacle Point seems to have been the brightest red available, and this, the authors argue, makes it unlikely that the ocher had a purely utilitarian function (Marean et al. 2007; McBrearty 2007). This example is not unique. Indeed, it is one of many early anticipations of so-called symbolic behavior. Thus body adornments, too, may not have been restricted to humans who have lived in the last 60,000 years or so. Even so, there is a consensus that human material culture, behavior, social life, and foraging economy changed profoundly between the time of the earliest members of our species and about 50,000 years ago. Perhaps for the first 100,000 to 150,000 years of our life as a species, human lives were quite unlike anything known from the ethnographic record.

Given this record, we face a puzzle of some kind. If we think that these First Sapiens were not cognitively like Moderns, what changed and why? What cognitive change explains the difference between truly ancient and more modern lifeways? If, on the other hand, we think that these ancient *sapiens* had essentially the same cognitive horsepower as those of the last 50,000 years, we have a different puzzle. Why did these humans take so long to generate the technology and social life that is such an evident and dramatic feature of the last 50,000 years? Why did it take so long for the ideological life of modern humans—art, symbols, the burial of the dead— to establish?[4] The question of ideology seems especially pressing, given both the centrality of such modes of thought to modern human life and the fact that they do not seem to depend on specific technological innovations. Burial of the dead, for example, is not technologically demanding.

## 3.2   The Symbolic Species

Thus we have something to explain; there is something problematic about the deep history of our species (and perhaps of its sibling species). But it is surprisingly difficult to precisely specify the phenomenon to be explained. Perhaps the dominating recent response is to focus on symbolic behavior, taking the use of physical symbols to be the breakthrough capacity (see, e.g., Henshilwood and Marean 2003; Wadley 2001; Zilhão 2007). Symbolic behavior in all its manifestations, from language to art, style, decoration, and ritual, seems genuinely central to what we are. The use of such symbols (the thought goes) shows that a new kind of social life has arrived. Moreover, one line of argument suggests that the use of public symbols is especially cognitively demanding, so that symbol use is an unequivocal sign of distinctive intellectual powers. So perhaps it is not surprising that archaeologists have come to focus on symbol use as the distinctive signature of the modern mind. Hence the excitement generated by João Zilhão's recent claim to have discovered indisputably indigenous Neanderthal shell jewelry (Zilhão et al. 2010).

Let me begin with the idea that the use of public symbols is central to behavioral modernity because such symbols have transformed human groups. One way that human groups differ from animal societies is that human groups are groups for themselves, not just groups in themselves (Cohen 1980). People belong to groups; they recognize themselves as a member of a group and often treat that fact as a central feature of their lives. Individuals identify with their communities and identity with their distinctive norms and customs. Culture, mediated by the use of symbols, welds people into members of a community who *identify* as members. Humans are not just members of communities; they *think* of themselves as members of communities (see, e.g., Chase 2007). Human groups are "symbolically marked"; they share distinctive norms, customs, rituals, and the like. Symbols are badges or insignia of group membership and identity. Mutual knowledge of these shared aspects of life underwrites individual identification with groups in which they are embedded. For those who think of culture in this way, the emergence of decoration, public art, and "style" is the archaeological signature of the transition from mere group membership to consciousness of membership (Henshilwood and Marean 2003; Wadley 2001). According to this view, the evolution of behavioral modernity is a cultural revolution, a transition from mere coexistence with others to identifying oneself with others. This transition is relatively recent; it took place (in this picture) somewhere between 120,000 and 50,000 years ago.

On this view, public symbol use is intrinsically important because it marks the arrival of a new kind of group. Rituals, customs, and norms are collective rather than individual phenomena, and often when archaeologists talk about the central role of symbol use in behavioral modernity, they have in mind the material traces of these group-defining activities. So, for example, McBrearty and Stringer clearly embrace this public and collective conception of a symbol when they write:

The ability to manipulate symbols is considered an essential part of modern human cognition and behaviour, although definite traces of symbols in the archaeological record are difficult to recognize and are often obscured by the ravages of time. All humans today express their social status and group identity through visual clues such as clothing, jewellery, cosmetics and hairstyle. Shell beads, and haematite used as pigment, show that this behaviour dates to 80,000 years ago in coastal North and South Africa. (McBrearty and Stringer 2007, 793)

According to this line of thought, symbol use is central to behavioral modernity because behaviorally modern humans are behaviorally modern by virtue of living in a distinctive kind of social world, a world of collectively made meaning. In turn, it is plausible that these new social worlds were the essential foundation of the elaborate, world-transforming cooperation and division of labor that is such a distinctive feature of our lineage.

So one reason for focusing on symbol use is the idea that a special connection holds between physical symbols and the distinctive form of contemporary social life. But it is also thought that symbol use shows a distinctive cognitive sophistication. Symbols are doubly important as markers that we—modern humans—are now on the scene. Although symbols depend on shared norms and conventions, individuals must be capable of understanding symbols and their use. One standard line of argument holds that symbol use requires an exceptionally sophisticated cognitive capacity. Hence we get two linked reasons for privileging public symbol use as a criterion of behavioral modernity: using public symbols defines a distinctively modern social world and demands a modern mind. However, these two reasons do not dovetail. One class of symbols may well define groups as self-conscious collectives. A second class of public symbols does indeed make heavy cognitive demands on their users. But these are different types of symbol. The "symbolic criterion" for behavioral modernity looks plausible when we conflate these two very different forms of symbol use, allowing us to think that symbol use is both cognitively sophisticated and definitive of a new form of cultural life.

So, for example, Lyn Wadley (2001) argues that symbolic cognition is the core feature of behavioral modernity, and in doing so, she relies on Terry Deacon's argument that the transition from icon to symbol in the evolution of language is a cognitive revolution, as the meaning of a symbol cannot be learned by an associationist mechanism (Deacon 1997). But Deacon's argument relies on two critical premises. First, words are arbitrary; no iconic or mimetic element supports interpretation, for (in almost all cases) no resemblance relationship holds between word and referent. Second, reference is (often) temporally and spatially displaced: we can and typically do use "tiger" to talk about tigers in their absence. Indeed, we talk about things that do not exist at all, for example, in fiction and myth.

However, symbols that serve as insignia of social place and identity are neither arbitrary nor displaced. Ocher markings, face paintings, and feathers and masks worn in ritual and ceremony are not temporally displaced. The ornaments a person wears to signify membership and status are on him or her. Likewise iconic elements almost certainly played a role in decoration and ornamentation—for example, in emphasizing or drawing attention to particular features. Nor is the significance of ornaments such as beads and ostrich shell fragments merely conventional. These seem to become common roughly 40,000 years ago (and to be a part of Neanderthal life too). But there are a few examples of early personal ornaments.[5] What should we make of these early African beads and the claims made about them? D'Errico and his coauthors write of 78,000-year-old beads from South Africa:

A key characteristic of all symbols is that their meaning is assigned by arbitrary, socially constructed conventions. . . . Personal ornaments and art are unquestioned expressions of symbolism that equate with modern human behaviour. (d'Errico et al. 2005, 4)

To the contrary. While these beads are clearly not utilitarian in any mundane sense, their meaning is not arbitrary. Their rarity suggests that they are special, and so they are most plausibly seen as expensive signals of status, skill, or success. They are Middle Stone Age Ferraris. And the whole point of owning a Ferrari is that its meaning is not conventional or arbitrary; its genuine cost means that it is an honest signal of success.

In short, understood one way, the ability of a mind to use and understand symbols really is a signature of cognitive sophistication. But those are not the symbols used in group self-identity and hence are not the symbols whose presence becomes obvious in the Upper Paleolithic and

Late Stone Age. In contrast, social marking is not, in and of itself, obviously a sign of distinctive cognitive capacities, capacities possessed only by the most recent, most large-brained members of our lineage. If the expansion of symbol use in Upper Paleolithic Europe, Late Stone Age Africa, and Holocene Australia is of central importance, it is because a new kind of human social organization had appeared, and the use of public signs of identity was both cause and effect of this change. It is not because the ability to use ocher or beads in itself signals a transition in individual cognitive sophistication.

## 3.3   Public Symbols and Social Worlds

Moreover, I doubt that the first regular appearance of physical symbols in the material remains of ancient peoples really does flag a transition to a new social world. The relationship between capacity, and the manifestation of that capacity in the archaeological record, is as complex for physical symbols as it is for other aspects of material culture. This point is sometimes overlooked in the literature. Wynn and Coolidge, for example, treat the appearance of complex, expensive grave goods in the European record of the Upper Paleolithic as evidence that Moderns were cognitively different from the Neanderthals they displaced. Some of these are spectacular: one grave contained teeth from at least 63 foxes; another contained over 5,000 beads (Wynn and Coolidge 2004, 480). No Neanderthal grave compares. But precisely because these sacrifices are so expensive, they may well signal increasing social hierarchy and wealth differential rather than new ways of thinking.

The production and use of physical symbols, as with other artifacts, depend on an interplay between demography, economics, and individual cognitive capacity. Consider, for example, recent arguments that insignia-symbols have a deep African history, 150,000 years or more, long predating the Upper Paleolithic (Conard 2006; d'Errico et al. 2005; Marean et al. 2007; Zilhão 2007). The most systematic of these early examples of symbolic behavior are burial of the dead and the use of ocher. But while we have evidence of fairly systematic burial of the dead, the significance of this practice is not clear. It is one thing not to treat *as refuse* the corpse of your father, sister, or daughter. It is another to construct a magical narrative about their ongoing significance. In the absence of grave goods, we have no evidence of magical narrative. In short, while burial of the dead is evidence of modern-like emotional attachment, it is not evidence of anything else.

More generally, the construction and use of physical symbols depends not just on capacity but on the nature of agents' environments. In this respect, the manufacture of physical symbols is like other material technologies. In an important discussion, Kuhn and Stiner (2007a) compare ocher and shell-based beads as signaling systems. Ocher significantly precedes the use of shells as beads in the archaeological record. Indeed, ocher seems to have been used widely in the Middle Stone Age, with deep dates. Some are very early Middle Stone Age, roughly 280,000 years ago (McBrearty and Stringer 2007). In part, this may be because ocher has uses other than in human-to-human signaling. Ocher may have purely utilitarian purposes: as a preservative, insect repellent, or ingredient of glue. But even if such mundane uses can be excluded, it does not follow that the use of ocher is symbolic, either in the sense of displaced reference or in the sense of social marking. It could, for example, be used in signal enhancement: making a face, shield, or person more visible, startling, or threatening. Imagine, for example, spooking animals by suddenly emerging from cover in a game drive. Signal enhancement would make such a tactic much more effective.[6] Camouflage is another possibility: for example, using ocher to break up contours. This suggestion seems especially relevant given recent reports of Neanderthal use of dark ochers. Neanderthals seem to have been ambush hunters; for them, camouflage would have been important.

So perhaps the use of ocher was established before humans regularly altered their bodies and garments to send signals to one another, and was then exapted as an existing technology to send signals. That might explain why shell beads, which have no use except as signals, arrive later in the record. But Kuhn and Stiner (2007a,b) show that ocher and shell-based beads have different properties as signals. As a rough model, think of eighteenth-century military uniforms and contrast the basic uniform itself with the symbols of rank and role on the uniform. As I read Kuhn and Stiner's discussion, shell-based systems are well suited for within-group signals. Shells can be standardized and compositionally organized. Their pattern and placement can itself be a signal, and one that can be duplicated or systematically varied: having, say, three rows of shells rather than two around one's neck can be a discrete, regular, and repeated signal. As a consequence, shell-bead systems can be like three crowns or two stripes on an army uniform; they have the capacity to encode precise information about rank, role, age, status, gender, or even individual identity (just as a sequence of color bands on a bird's leg can be used by ornithologists to identify individual birds). Such signaling systems are precise and rich, but they such have low amplitude. The precise pattern is difficult to see at any distance (and hence ornithologists must typically equip themselves with high-quality optics to read their own signs). Moreover, if the comparison with symbols of rank or identity is apt, the system is both somewhat complex and arbitrary. The significance of a particular array will be obvious only to insiders. In contrast, ocher has high amplitude.[7] As with eighteenth-century military uniforms, a shield, face, or garment colored in a distinctive way is recognizable at a distance. In contrast to shell-bead signals, ocher-based signals would be well designed to signal group membership or identity to another group. And while such a signal is arbitrary, it is not part of a complex system, nor is it displaced in space and time from its target. Identity symbols could be learned simply and quickly by individuals in other groups.

If these considerations are persuasive, the appearance of ocher, beads, and the like in the archaeological record is an effect of demographic and social change, not the first emergence of self-identifying groups.[8] In simple social environments, these social signaling systems would be unnecessary: there are no strangers to read signals. This demographic suggestion is supported by the fact that physical symbol making emerges at different times in differing *sapiens* groups. Symbolic marking is pervasive in Upper Paleolithic Europe about 35,000 years ago; it is not pervasive in Australia until roughly 5,000 years ago (Brumm and Moore 2005). According to the population-structure hypothesis, the appearance of physical symbols of group membership in the archaeological record has nothing to do with people first beginning to think of themselves as members of groups. Rather, it is the invention of advertising. Members of a group only needed to badge their identity once their social world became dense enough. After that threshold, they regularly met others who did not know them as individuals located in a specific network. That transition selected for physically advertising group membership (Brumm and Moore 2005; Kuhn et al. 2001; Kuhn and Stiner 2007a). Beads and similar low-amplitude, short-range signals appear later as groups become more internally complex, more differentiated, and perhaps more hierarchical. Kuhn and Stiner's specific hypothesis is quite plausible, but more importantly, it illustrates a more general idea. Physical symbols have particular social roles, and hence such symbols will be much more archaeologically visible in some social contexts than others. Physical symbols are just one fallible indicator of cultural richness. We have no smoking gun, no material trace left when and only when agents live in social worlds that fall within the modern range.

## 3.4  Preserving and Expanding Information

The focus on symbol use as a criterion of behavioral modernity is a symptom of a more general problem: that of thinking of behavioral modernity in terms of the capacities of individual agents. I suggest that it is much more productive to think of behavioral modernity as the effect of a favorable feedback loop between individual cognitive capacity and the social and informational environment. In particular, we see behavioral modernity when the capacity to preserve informational resources is augmented with a reliable capacity to incrementally improve those resources and transmit the improved package to the next generation. In this regard, ancient Australia provides an instructive example.

Ancient Australia shows the problems inherent in thinking of behavioral modernity as a feature of individual agents. Almost everyone accepts that the First Australians must have been behaviorally modern, because of the challenges they overcame to reach the Sahul. They must have been able to plan, coordinate, and innovate. The initial expansion of humans into the Sahul about 45,000 years ago could not have been accidental (Allen and O'Connell 2008). The sea crossings are too formidable for humans to have arrived through any scenario resembling a "pregnant women on a log."[9] However, before the Last Glacial Maximum, 20,000 years or so ago, the archaeological record resembles that of Middle Stone Age Africa.[10] For the first 25,000 years they were here, the first Australians seem to have possessed a limited technological toolkit, exploited a narrow resource band, and showed limited signs of symbolic culture. Eventually the standard symptoms of behavioral modernity do appear in the Australian record. But as in the African case, the supposed behavioral and archaeological signatures of behavioral modernity do not appear together in space and time, and the stone toolkit stays quite simple until the Holocene (see Habgood and Franklin 2008, 211, fig. 9). Only over the last 20,000 years do we consistently see the usual archaeological signatures of behavioral modernity: broad-range foraging, environmental management, technological innovation, and obvious symbolic culture (Brumm and Moore 2005; Keen 2006; O'Connell and Allen 2007).

Jim Allen and James O'Connell (2007) interpret this record as showing that people can *be* behaviorally modern without *showing* that they are behaviorally modern. As a consequence of environmental and demographic factors, modernity left no trace for upward of 25,000 years. O'Connell and Allen do not consider the idea that Australians *ceased* to be modern after they arrived. Neither do Habgood and Franklin. Neglecting this possibility makes sense if we think that modernity is coded and canalized in individual genomes, if it is an attribute that individuals have largely independently of their cultural environment. But it makes no sense if behavioral modernity is wholly or partially constituted by the organization of social life. That might have changed fundamentally as small numbers of people dispersed into an enormous landscape. The communal resources available to very small groups dispersed over enormous and inhospitable distances would be very different from those available to communities based on the fertile islands and shallow seas of Southeast Asia. Social learning is more robust when there are many models rather than few. The potential to lose behavioral modernity will be important in the next section, when we discuss Neanderthal extinction.

I suggest that behavioral modernity itself is the collective capacity to retain and upgrade rich systems of information and technique. That capacity depends on the interaction of individual minds, organized learning environment, and population structure. In this view, the specific features of the archaeological record (symbol use, composite toolmaking, ecological breadth, and the like) are just fallible indicators of this basic cognitive-cum-cultural capacity. None have special significance in themselves. But they are signs of a threshold. We see the signal of modernity when a stabilized system of interaction makes the accumulation of cognitive capital reliable. We can draw an important distinction between the conditions that allow information to be *preserved* reliably and those that allow it to be *expanded* reliably. So the final element needed for behavioral modernity was the reliable preservation of incremental improvement, allowing cultural learning in partnership with trial-and-error learning to become a mechanism of the cumulative growth of informational resources.

The distinction between preservation and expansion allows us to make sense of the hominin record. That record seems to show three phases: a long phase of preservation, with at best very slow improvement, a not yet stable shift to expansion, and a final phase in which innovations and additions to the communal stock of information are much more reliably transmitted to the next generation (though still with some possibility of breakdown). Thus hominin history began with an extremely long phase of technological conservatism. Simple chopping tools and flakes emerge approximately 2.6 million years ago in Africa.[11] At about 1.8 million years ago, this technology is supplemented with the classic Acheulian hand ax (and perhaps also with worked bone tools [Backwell and d'Errico 2008]). These hand axes are bifacially flaked and often have a somewhat standardized teardrop shape, perhaps more so late in the period. The extreme conservatism of the products of stone technology probably masks a greater rate of innovation in technique and in other aspects of life. Stout (2001) argues that there was incremental improvement in stone shaping techniques, especially by about 800,000 years ago. Moreover, at some stage in this period, fire was domesticated, cooking became important, and there were major changes in diet (probably involving both meat and expanded plant exploitation). Brain, body, and life history all moved toward the late hominin pattern in this period. So there was some innovation, though very slow innovation.[12] Middle Stone Age points begin to appear about 280,000 years ago. It is natural to interpret this as the origin of composite tools, for points need to be attached to a shaft. Moreover, the points themselves require a two-step manufacturing process. So sometime between 300,000 and 200,000 years ago, but probably closer to the second date, technological and ecological traditions become less conservative. As we have seen, innovations in this period anticipate later technological revolutions, but often these innovations seem to fade out. The accumulation of innovation is not yet stable. The final phase, of course, is the signature period of behavioral modernity: innovation, regional variation, and expansion into all but the most forbidding regions.

From this sketch, we can see that hominin history has two large-scale features that our model of modernity must capture. One striking element of this pattern is that innovation is initially very slow. This is the phase of (mostly) mere preservation: core skills can be preserved. No doubt innovations occurred, but they established only by lucky accident; they had to induce a dramatic-enough change in a group's way of life for by-product social learning to be reliable, and remain reliable, over ecological fluctuations. Perhaps the domestication of fire might be such a case. A second is that change in technological competence is not unidirectional (and this is especially so over the last 300,000 years). Technological and ecological innovations appear (and establish over sufficient space and time to leave a trace) but then disappear again. In their reviews, Conard, d'Errico and Stringer, and Hiscock, and O'Connor all emphasize these early appearances of technologies that become signatures of later periods (Conard 2006, 2007; Hiscock and O'Connor 2006; d'Errico and Stringer 2011). For example, microliths are regularly shaped, pointlike artifacts that are often taken to be a signature technology of behavioral modernity, both because they can be made in regionally distinctive but still regular ways and because they are thought to have been mounted on spears or arrows. Hence they show the capacity to make multipart artifacts. Yet Hiscock and O'Connor point out that microliths are found in significant numbers early in one region of the African Middle Stone Age (dated to between 300,000 and 250,000 years ago) and again late in the Middle Stone Age (as part of the Howiesons Poort industry, perhaps 70,000 years ago). So microliths are found before the establishment of paradigm behaviorally modern cultures, but only patchily in space and time. Here we have both preservation and expansion, but less reliably. Innovations are easily lost.

So perhaps sometime between 300,000 and 150,000 years ago, hominin culture became cumulative in two senses. The volume of culturally mediated learning increased: a larger range of hominin action owed its character to intergenerational social learning. Thus the range of materials expanded (including ocher, bone, antler, and ivory). The variety of tools increased, in part because technology took on new functions. It was used to make material symbols and other artifacts, to build shelters and organize domestic space (Wadley 2001), and to make clothing, using awls and needles (Gilligan 2007). Hominins expanded the range of resources they exploited (O'Connell 2006). Moreover, at some point in hominin evolution, children came to learn the norms and customs of their community, not just the local techniques for making a living. Human behavior became more diverse and less stereotyped, in ways that were guided by information flows from the preceding generation. The bandwidth of cultural learning expanded. But cultural transmission gradually became cumulative in a second sense as well, permitting the stepwise improvement of specific technologies. For example, fire was almost certainly domesticated in stages, beginning with the maintenance and exploitation of natural fire, probably followed by the development of techniques for making fire portable (the dates of the first use of fire are decidedly controversial, but we have clear signals from about 800,000 years ago [Alperson-Afil, Richter, and Goren-Inbar 2007; Goren-Inbar 2011]). These important breakthroughs were followed by ignition technologies and improvements in the control and use of established fire, in hearths and the like (Brown et al. 2009; Ofek 2001). Stepwise improvement requires high-fidelity transmission. In the hominin record, the expansion of bandwidth seems to be roughly correlated with increasing fidelity (assuming that more complex technologies depend on higher-fidelity learning). Behaviorally modern cultures depend on both high fidelity and expanded bandwidth. So we need an explanation of both aspects of cultural accumulation.

Thus cultural transmission has had an increasing footprint in the hominin record, a trend culminating in behaviorally modern cultures, with stable, high-volume, high-fidelity information flow across the generations. In my view, stable, high-volume, high-fidelity flow depends on three factors: individual cognitive adaptation, adapted learning environments, and demographic support. The most obvious of these three factors is the evolution of minds adapted for cultural learning and teaching. No one doubts that we have minds adapted for social learning, and no one doubts that those adaptations play a central role in human life. Humans have advanced theory of mind, powerful communicative capacities, and a talent for observational learning. Often humans have enough reflective understanding of their own skills to teach them effectively. That is important, for the skill sets of behaviorally modern humans are so elaborate that teaching is probably essential. Second, as I argued in section 2.3, high-volume, high-fidelity flow depends on organized learning environments. Trial-and-error learning is socially supported and enhanced: adults often organize learning environments to enhance individual adaptations for social learning. I will not replay those arguments here but instead turn to a third important precondition of stable, high-fidelity, high-volume flow: demographic support.

Group size matters because larger groups can afford to support specialization; smaller groups cannot. Thus Haim Ofek (2001) has noted that a larger market size allows more specialization and more division of labor, both of which impact positively on a group's informational resources. A small group will not be able to afford a specialist fire keeper or bow maker; a medium-sized or large group perhaps can. They will have enough customers to support specialization. Specialists typically have higher skill levels and hence set a higher bar for the next generation. Moreover, a more diverse group with a varied skill set is more likely to innovate than a small, more homogeneous group. Those who specialize in a craft are the most likely to find an improvement in it, and innovation through cross-fertilization is more likely as the overall skill base becomes more diverse and extensive. Specialists may also be more accurate in filtering unsuccessful innovation, and as Magnus Enquist and Stefano Ghirlanda (2007) show, filtering is essential if culture is to become cumulative.

Probably more important, though, redundancy plays a critical role in buffering the group's informational resources. Larger groups store information in more heads than smaller groups. Information can easily drift out of a small group, through unlucky accidents to those with rare skills. The fewer individuals who have a particular skill, the more easily it is lost by unfortunate accident. If high-fidelity transmission depends on redundancy, and on the availability of many models, rarely used skills will be lost, and frequently used skills will be transmitted with less fidelity. More generally, to the extent that redundancy is important to the retention of information, smaller groups will retain information less well than larger groups. Impressive recent modeling work (connected to suggestive case studies) shows that information loss through unlucky accident is likely to have a dramatic and brutal impact on small populations, especially if they are not well connected to other groups.[13] These models make a convincing case for the importance of demography. Moreover, Powell, Shennan, and Thomas's (2009) extension of Henrich's work shows that the models are robust, and that the parameter values that predict accumulation map quite plausibly onto estimates of human populations just before the establishment of behavioral modernity.

In addition, redundancy may play a second role in compensating for low-fidelity cultural learning. Modern humans are clearly individually adapted for social learning (Csibra and Gergely 2005; Gergely, Egyed, and Kiraly 2007; Tomasello et al. 2005). But Richerson, Boyd, and their colleagues doubt that these adaptations suffice for high-fidelity learning, and argue that the social environment compensates for low fidelity through redundancy. Naive agents have many opportunities to acquire specific skills and critical information, and they develop models to show that redundancy—for example, a naive agent using many models rather than a single model—can compensate for low-fidelity one-on-one learning. With sufficient redundancy, and with members of a population connected in the right ways, a population can preserve its informational resources in transmission to the next generation through low-fidelity channels (Gil-White 2005; Henrich and Boyd 2002; Henrich, Boyd, and Richerson 2008; Richerson and Boyd 2005).[14]

In summary, then, the cultural learning characteristic of the Upper Paleolithic transition and later periods of human culture—social transmission with both a large bandwidth and sufficient accuracy for incremental improvement—requires individual cognitive adaptations for cultural learning, highly structured learning environments, and population structures that both buffer existing resources effectively and support enough specialization to generate a supply of innovation. Middle Stone Age instability—the appearance and loss of innovation—hints at a failure of one of these supports. Loss of innovation is no surprise if these hominins lived in small bands, largely isolated from one another. Likewise it is no surprise if Middle Stone Age learning environments were less rich; if, for example, skills largely depended on automatized habit, as Wynn and Coolidge (2004) believe. It is one thing to be able to construct a hafted spear by using a heated mix of ocher and acacia gum to glue the warhead to the shaft. It is another to be able to demonstrate that skill in ways that make the most crucial elements salient and transparent. If Middle Stone Age hunting was less efficient, generating a smaller surplus, adolescent learning would be truncated. The young would need to learn their trades more rapidly, and that too would make innovation more apt to be lost.

Information-rich, expertise-dependent forager lifestyles depended on a combination of an organized learning environment and specific adaptations for social learning. The cultural and technological innovation dramatically visible in the archaeological record in the Upper Paleolithic Revolution is a signal of such a social world: a social world that supports high-fidelity, high-bandwidth transmission across the generations. Individuals in these social worlds were cognitively equipped for social learning. But they also depended on an adapted environment and on populations that spread risk and supported specialist expertise. The persistence of these lifeways depended on highly skilled agents sharing their expertise and on the reliable replication of the learning environment in which crucial expertise was acquired. This combination, and only this combination, allowed cognitive capital to be accumulated and behaviorally modern cultures to emerge. These social worlds are stable and flexible and can be adapted to very different social environments, ecological demands, and forms of expertise, as section 2.3 documents. But their stability has limits. Neanderthal Europe may illustrate how these mechanisms of cultural adaptation and transmission can fail.

## 3.5   Niche Construction and Neanderthal Extinction

An important strand of work on Neanderthal extinction presupposes that Neanderthals were displaced by *sapiens* through some important sociocultural difference between Neanderthal and *sapiens* groups. In turn, this sociocultural difference was due to an intrinsic cognitive difference between *sapiens* and Neanderthal minds. For example, Steven Mithen and others have argued that the Neanderthals lacked full human language (Bickerton 2002; Lieberman 1998; Mithen 2005; Tattersall 2009). It is indeed possible that there were significant cognitive differences between Moderns and Neanderthals. But the extinction of the Neanderthals does not show that they differed cognitively in important ways from us. Rather, I think it is likely that they were pushed over the brink by negative feedback loops between demography and environment. These loops eroded the preconditions of high-volume social learning. Just as Moderns illustrate the effects of crossing a positive threshold, Neanderthals may well illustrate the effects of crossing a negative threshold, after which they could not retain the informational resources needed for survival.

The evidence seems to show that beginning around 50,000 years ago, the Neanderthals were a species under stress. They had already disappeared from large parts of their former range. Over the next 20,000 years, they dwindled further in range and numbers, with the last surviving populations scattered across the Iberian Peninsula. Those too disappeared over the next few thousand years as local extinction became final extinction. Explaining Neanderthal extinction has generated deep controversy within paleoanthropology. According to one view, Neanderthals were driven to extinction by the invasion of behaviorally modern *sapiens* (O'Connell 2006). The competitive superiority of the invaders was usually supposed to result from intrinsic cognitive differences between *sapiens* and Neanderthals (superior language, superior capacity to integrate information across multiple domains, superior capacities to innovate, superior working memory). An alternative idea is that the Neanderthals were driven to extinction not by invasion but by unfavorable environmental changes.[15]

I take Clive Finlayson as my standard-bearer for the idea that Neanderthal extinction was environmentally forced. That picture can be made plausible. For one thing, as Finlayson points out, we have little direct evidence that Neanderthals and Moderns were competitors, that they occupied the same region at the same time.[16] Moreover, Neanderthals were adapted, both biologically and culturally, to a cool, temperate world. They were physiologically adapted to cool climatic conditions: they were robust, with head and body adaptations somewhat similar to, but probably more extreme than, Arctic-adapted Moderns. Perhaps more crucially, their foraging economy (together with their short-armed, robust physiques) was adapted to ambush hunting of large game. The classic Neanderthals were not pursuit-endurance hunters; they depended on cover to get close to game. The cool, temperate, forested world to which they were adapted was in retreat for much of the period between 50,000 and 30,000 years ago—the period in which they dwindled to remnant, vulnerable populations. Thus Finlayson argues that the Neanderthals were in deep trouble whether or not Moderns exerted pressure on Neanderthal populations. Indeed, he argues that climate changed doomed them.

By the time the classic Neanderthal had emerged, during the last interglacial around 125 thousand years ago, they were already a people doomed to extinction. Like the hippos, rhinos, and elephants of the Eurasian forest, the Neanderthals were a population of living dead, existing on borrowed time. Like other mammals, the Neanderthals had a short-lived moment of reprieve as the climate warmed up. The next time the climate would be this generous, 100 thousand years later, Neanderthals, straight-tusked elephants, and narrow-nosed rhinos would no longer be. (Finlayson 2009, 116)

Finlayson, like Robert Foley (Foley and Gamble 2009; Foley and Lee 1989), resists human exceptionalism. The factors that explain the distribution and abundance of humans are those that explain the distribution and abundance of other large mammals. Thus Finlayson argues that the glacial-interglacial cycle accelerated and intensified, with deeper glacial troughs, and this drove many temperate-adapted large mammals to extinction. In these intense glacial troughs, warmer refuges were too small or too ephemeral to support robust populations. Small populations of large, slow-breeding mammals are always vulnerable to extinction. So even if warmer habitat islands persisted through the deepest glacial troughs, it comes as no surprise that these cycles eventually drove interglacial specialists to extinction. According to this view, we have no special problem of explaining Neanderthal extinction. The Neanderthals were just one of a guild of temperate forest and woodland specialists that all went extinct as their world disappeared around them.

I am skeptical: I do not think that it is legitimate to treat Neanderthal extinction as just another example of temperate-adapted mammals failing to survive unfavorable change. For humans, adaptation has often been accelerated by cultural innovation. Moreover, humans adapt their environment to their own phenotype with fire, shelter, and clothes. Neanderthals were intelligent, technologically adept, niche-building humans; comparing their extinction to that of the narrow-nosed rhino will not do. Thus I agree that once Neanderthals were confined to small, relict populations, they were indeed extraordinarily vulnerable: their encultured lifeway became a risk factor rather than an escape clause. But we also need to explain why Neanderthals were pushed into refugia, given the human capacity to respond adaptively to novelty. I begin with the idea that refuges are traps rather than shelters.

*Life in a refuge.*   Once confined to small, isolated populations, the Neanderthals would have been at even greater risk of extinction than, say, an island population of dwarf woolly mammoths. Culturally based adaptations are fragile in small populations. First, the economics of hunting become less reliable as group size falls. Large-game hunting has a variable rate of return. When hunts are successful, they yield large rewards. But often they fail. What then? As a general rule, the Neanderthals do not seem to have compensated for variance in hunting success by the gendered division of labor typical of recent *sapiens* forager societies, in which women concentrate on small game and plant-based food. These yield lower average returns, but they are more certain (Kuhn and Stiner 2006). How, then, might Neanderthals compensate for variable returns? In a colder climate than the subtropical heartlands of Moderns, storage might help. So too, in all probability, would sharing between bands and by a single, larger group supporting multiple simultaneous hunts. Both insurance mechanisms become less reliable as numbers decline. If islands of survivability are small and support only small bands, the capacity to sustain simultaneous, complementary hunting parties will decline. So too will mutual aid between groups. Stanley Ambrose argues that intercommunal support networks—sustained by regular visits, gift giving, and intermarriage—are crucial for forager survival in extreme environments.[17] Such intercommunal networks are more fragile and less effective as populations shrink and become confined to scattered favorable islands (Ambrose 2010).

Indeed, at the limit, a small group might lose its ability to hunt large game effectively through an inability to muster enough adult males to ambush effectively. Neanderthals were social predators, and the more social the predator, the more vulnerable it is when living in small numbers. Obviously we cannot reconstruct the social organization and ecological dynamics of Neanderthal hunting, so we do not know the optimal size of a Neanderthal hunting band. Ideal band size will depend on terrain, target, cover, and much else. But they hunted large herbivores at close range by using stabbing spears. So we can reasonably conjecture that relatively safe, relatively effective hunting required hunting teams that were closer in number to ten than to two. Paul Bingham (1999, 2000) has emphasized the importance of coalition size in making an attacking coalition deadly to its target and relatively safe for coalition members. He developed his model for human-on-human violence, but the model extrapolates to the violence of hunting.[18] So long as each member of the coalition has room to attack, a larger coalition is safer than a smaller one, because return fire is distributed over more targets (reducing the danger to each), and because the target can return fire for a shorter period. The higher volume of incoming fire kills, disables, or forces into flight the target more quickly. So long as the band can coordinate and attack simultaneously, a larger team is safer and more effective than a smaller one. As band size declines past a crucial threshold, hunting becomes more dangerous and less effective.

Thus the economics of the Neanderthal way of life would have been subject to increasing stress the more Neanderthals were forced into small groups that could not support one another. Similarly, declining group size would have a negative effect on a group's ability to retain its skill base.[19] The central claim of my whole work is that the distinctive character of human social life depends on the accumulation, preservation, and intergenerational transmission of cognitive capital. Neanderthal lifeways depend on extensive cognitive capital. Ambush hunting depends on natural-history expertise and skills in stalking and concealment, and it requires weapons whose production requires high levels of skill. Neanderthal material culture was not highly diverse (except perhaps in its latest manifestation). But stoned-tipped spears are composite tools. They are hafted weapons, with the spearhead attached to the shaft with birch resin.[20] Moreover, while their stone technology is not highly diverse, the manufacturing technique is complex and difficult, as Wynn and Coolidge (2004) emphasize in their review of the supposed cognitive differences between *sapiens* and Neanderthals. Earlier Neanderthal stone tools are made using Levallois techniques, in which the knapper first prepares a shaped core with a convex striking surface. That striking surface is then itself shaped, so that force is transmitted through it to a second surface. This second surface is the zone from which flakes are struck for further shaping and sharpening, by hard hammering on the striking zone. So Levallois expertise depends on the capacity to choose raw materials of the right kind and plan a complex sequence of transformations—a sequence that will vary from case to case, depending on the shape and quality of the initial stone and on the target tool. Stone knapping also depends on complex motor skills: shaping the core with a striking platform and flake source and then striking the platform at the right angle and intensity to generate sharp flakes. Somewhat later, Neanderthal toolmaking tends to shift to a "Quina" system, more akin to Upper Paleolithic methods (Hiscock et al. 2009). So while Neanderthal methods were not invariant, they were skilled. Modern re-creations of stone tools suggest that these are difficult skills, and "mastery emerges only after years of practice" (Wynn and Coolidge 2004, 475).

As we saw in section 3.4, cognitive capital is hard to retain in small, scattered populations under stress. Under these conditions, information is not buffered by redundancy, and specialization is not protected by market size. If survival depends on the cultural transmission of large amounts of information and expertise, the chance of survival goes down as group size declines. Such transmission is less robust, more fragile and noisy, as groups become small and isolated. These demographic factors will be exacerbated if the foraging economy is under stress. Michael Gurven and Kim Hill argue that the human life history pattern depends on extensive intergenerational resource transfer: children and adolescents are supported (wholly or in part) for many years while they acquire the full forager skill set. This life history pattern is stable only because first, adults (especially adult males) generate a large surplus, and second, adults have low mortality. Investment in their expertise pays because they live long enough, on average, to take full advantage of their skills (Gurven and Hill 2009; Robson and Kaplan 2003). If Neanderthal populations were under stress, in less suitable habitats, and hunting under less favorable circumstances, the stability conditions of the human life history pattern may have been compromised. If, for example, hunting teams were too small, increasing the risk in hunts and pushing down adult life expectancy, adolescents may well have been pushed into the front line before they had the necessary skills and physique—a vicious cycle waiting to happen. In short, as Neanderthal populations became small and scattered, that in itself threatened their capacity to retain crucial expertise, even if all those fragments lived in ideal habitats (Richerson, forthcoming; Henrich 2004; Powell, Shennan, and Thomas 2009). If the fragments are less than ideal, the capacity to retain information is further compromised, both because adults are less able to support adolescents through long apprenticeships and because adults are more likely to die and disappear as models and as sources of advice and protection.

*Retreat or adapt?* Thus, if climate change drove Neanderthal populations down, fragmented them, and pushed them into isolated refuges, then the species was indeed doomed, and doomed because of its dependence on culture and cooperation, rather than in spite of its dependence on culture and cooperation. But why did the Neanderthals retreat? They were not narrow-nosed rhino, with adaptive suites that change only on the time scales of genetic evolution. The question of Neanderthal retreat has led to much speculation about the cognitive limits of Neanderthals, for it is culturally and technologically possible for foraging societies to make a living in cold steppe-tundra environments. These environments support significant herbivore-grazer populations and these can be exploited as the sustaining resource of forager lifeways. Despite speculation about cognitive limits (Mellars 2005; Mithen 2005; Wynn and Coolidge 2004), it is unlikely that Neanderthals were cognitively incapable of using the technologies that make cold-climate survival possible. Their core technology was sophisticated. Moreover, a late Neanderthal culture, the Châtelperronian, was a transitional technology somewhat akin to the Upper Paleolithic technology of Moderns (Finlayson and Carrión 2007). It incorporated many of the technological elements needed to exploit the steppe-tundra grazers (for example, tools for making the fitted clothes essential for surviving in truly cold conditions) (Gilligan 2007). Finally, Finlayson's data on Neanderthals living near Gibraltar very near the time of final extinction show that they had a broad range of subsistence strategies, though even these Neanderthals mainly targeted large and medium game (ibex and red deer). So instead of retreating, the Neanderthals could have adapted to the colder conditions and the resources they offered.[21] Why did they fail to adapt to a colder world? After all, we know that it was possible for foraging people to survive in that colder world.

The Neanderthals, I suspect, were caught in a fitness trap. They were adapted to a sinking local optimum, ambush hunting. This local fitness peak was separated from higher optima by a fitness trench. To survive in the steppe-tundra, they would have to switch from ambush hunting to pursuit-endurance hunting. That switch would have extremely high transition costs. (i) The steppe-tundra grazers (reindeer, musk oxen, and the like) are a distinct fauna to that of the temperate forests, so information about natural history and habitat, tracking, target identification, and target behavior would all have to be relearned. (ii) Likewise, the weapons skills are different. Exploiting the tundra-steppe grazers depends on projectile weapons rather than heavy, short-range thrusting spears.[22] Throwing a javelin with power and accuracy is a new and difficult skill (perhaps more difficult for Neanderthals than *sapiens*, given Neanderthals' squatter body shapes). (iii) The transition would require a range of new technologies. Most obviously it required new and lighter weapons. But it would require advanced clothing—properly cut, fitted, and sewn, rather than the cloaklike wraps that probably sufficed in temperate climates (Gilligan 2007). Moreover, pursuit hunting required the foragers to follow the herds, and this in turn probably required them to construct shelters rather than rely on natural shelters (caves and the like) in a well-explored home range (though perhaps they already had these skills. Peter Hiscock informs me that some Neanderthal cave sites were protected by shelter constructions). (iv) At least initially, the learning costs would be high, for there would be few, or no, locally available models. Those making the transition would have to invent many, perhaps all, of the skills and technology for themselves. If Neanderthals were confronted with a changing world after first contact with Moderns, there may have been a few artifact templates available. The same is true if some Neanderthal groups lived in proximity to other, adapting

Neanderthal communities. Some innovation might spread horizontally from group to group. But even if there was technological diffusion, Neanderthals would have had few opportunities to learn new techniques by observation and advice.

Even under favorable conditions, with rich social support, complex skills are acquired only by years of persistent efforts. Initial Neanderthal adaptation to steppe-tundra hunting could not be acquired under such favorable conditions. To adapt to a new world, the Neanderthals needed to embrace a technological, behavioral, and social revolution. The costs and risks of such a revolution would be extremely high. At any given moment, it would often have seemed a better bet to track the retreating habitat, relying on the skills and tools already mastered. The alternative was to chance a risky transition to a new lifeway. No doubt some groups did risk that change, as the Châtelperronian shows. But the cost of change may explain why the dominating response of Neanderthals was to track the conditions to which they were already adapted, even though those habitats were shrinking, rather than risk massive change. That is even less surprising once we realize that a shift to pursuit hunting could not be an individual-by-individual decision. A potential hunting party would have to decide to roll the dice, and so to the ecological costs we need to add those of coordination and persuasion.

The fitness-trap effect may have been exacerbated by two other factors. Adapting to the new conditions would have been much more difficult if the change from ambush hunting under forest cover to pursuit hunting in open grasslands was geographically or temporally abrupt. Since adaptation cannot be instant, and as foraging peoples do not have large stored surpluses, foragers must be able to make some kind of living using their old skills while extending those skills and acquiring new ones. In Africa, the transitions from forest to woodland to savanna are often gradual, with many hybrid zones. More open woodlands and shrublands connect true forest to genuinely open grasslands. In such circumstances, foragers can ply one trade while learning others, and the effectiveness of a technique—say, stalking using cover—will decline slowly as cover becomes harder to find. Making a living in the forest while learning about shrubland resources would be much more difficult in the face of knifelike transitions from, say, closed forest to open grassland. When the predominant vegetation depends on interactions between altitude and temperature, ecological transitions can be abrupt indeed. On mountains, the transition from forest to open grassland often really is knifelike. In the European glacial cycles, the transition from temperate forest to steppe or tundra may often have been like this, denying Neanderthals the opportunity to reduce learning costs by exploring mixed strategies of experimenting with and improving pursuit skills while still profiting from ambush. To pursue, they may have had to follow herds well away from surviving forest patches, well away from the opportunity to ambush in areas they knew well. A similar problem arises if the climate changes rapidly and intensely over time, for example, if a forest becomes a steppe in a generation or two. Then adaptation depends again on paying the full, and probably unaffordable, costs of trial-and-error learning. Both Gilligan (2007) and Finlayson (2009) suggest that, at least in some places, change was brutally rapid and intense.

Finally we need to consider the role of Moderns. We have little hard data on the timing and direction of expansion by *sapiens* into Eurasia or of competitive interactions between *sapiens* and Neanderthals. Certainly we find no evidence of competitive exclusion of Neanderthals by *sapiens* in the Levant between 100,000 and 50,000 years ago. The picture there looks more like oscillation and replacement, with Neanderthals expanding south in cooler periods, and *sapiens* expanding north in warmer periods, but not expanding into southeastern Europe proper, through Neanderthal habitat. This suggests that while conditions remained temperate and wooded, and ambush hunting remained a viable foraging strategy, Neanderthals excluded *sapiens*, at least before *sapiens* were fully behaviorally modern (Shea 2003). But somewhere between 50,000 and 40,000 years ago, *sapiens* added cultural adaptations to pursuit and endurance hunting to their existing physiological adaptations to that strategy, shifting to the use of projectile weapons (javelins, perhaps with woomeras) (Shea 2009). Plains hunting preadapted them for tundra-steppe hunting. Perhaps Moderns moved in with the shift to a cold-climate grazing world. If so, Neanderthals would have had to adapt both to a new climatic regime and to the presence of a competitor already adapted to the key demands of pursuit hunting. The Neanderthals may well have been caught in an exacerbated fitness trap: a trap exacerbated by the abruptness of change and the presence of a preadapted competitor. If so, we certainly do not have to posit Neanderthal cognitive inferiority to explain why they retreated with their favored habitats. But once they did retreat, critical cultural adaptations came under increasing stress, and in environments that were extremely unforgiving, even in refuges. Just as the increasing elaboration of *sapiens'* informational resources and material culture depended on feedback, so too did the Neanderthal sink toward extinction.

# 4 The Human Cooperation Syndrome

## 4.1 Triggering Cooperation

The argument of this book is that hominins increasingly diverged from their great ape relatives as their capacities for cooperation, information-guided foraging, and niche construction coevolved. Cognition and cooperation fed off each other as investment in cognitive capacity gave rich returns in a cooperative world. Think, for example, of the impact of increasing tolerance on social learning. Tolerant social worlds allow the young access to many models. Moreover, these models are more valuable because, with tolerance, close attention to adult activity is rewarded rather than punished. Exploration is more rewarding and less dangerous. Once these positive feedback loops are established, they drive divergence. But this picture does presuppose some initial trigger. If this coevolutionary picture is right, there must have been some environmental difference between early hominins and their great ape relatives, a difference that, perhaps in conjunction with some relatively minor phenotypic difference, initiated a diverging trajectory. One possibility is increasing climatic variation. Increasing climatic variation plays an important role in the work of Kristin Hawkes and her collaborators. Their focus is on the role of reproductive cooperation in human evolution. They emphasize its importance, arguing that hominins began to resemble humans with the evolution, about 1.8 million years ago, of *Homo erectus* (and perhaps its sibling species). It was with these hominins that reproductive cooperation, and hence cooperation in general, became central to our lineage. These authors take the increasing seasonality of hominin habitats to be the external trigger that pushed our lineage toward cooperative solutions to environmental challenges.

Hawkes and her colleagues think that seasonality selected for reproductive cooperation for feeding young in the hungry season (see, e.g., O'Connell, Hawkes, and Blurton Jones 1999). But increased seasonality selected for ecological as well as reproductive cooperation. Pleistocene Africa boasted an impressive stock of predators, and increased seasonality turned hominin habitat into savanna and woodland with much less natural cover than great ape forest habitats. Moreover, early hominins were not physically imposing. Thus selection for effective response to predators would have been strong. Moreover, Richard Wrangham (2009) points out that increased exposure to predators while foraging in more open habitats was coupled with increased exposure to predators while resting and sleeping at night. Habilines and erectines were probably incapable of sleeping in relative safety in nests in trees, as chimps do. Finally, collective defense is a relatively undemanding form of cooperation. To understand this, contrast collective defense with collective hunting. In many circumstances, and with many targets, a hunting party poses a much more dangerous threat to its target than does an individual hunter. But for a hunting party to be effective, its members must both coordinate with one another and suppress free riding. Hunting involves effort, and often risk, so some may be tempted to let others pay the costs. Then, once hunting has been successful, a further problem arises. The spoils must be divided, and that creates a potential flash point, a potential site of conflict that would undermine any further cooperation. Collective defense likewise depends on coordination and the prevention of free riding. But there is no extra problem of dividing the spoils. When collective defense has succeeded, the profit—safety—is automatically distributed to the defending coalition. There is no danger of the coalition imploding as it squabbles over the spoils of victory.

So while reproductive and informational cooperation had early origins, I conjecture that collective defense, perhaps using thrown stones (Bingham 2000), was likewise an early and important element of hominin cooperation. Furthermore, successful collective defense preadapts for further cooperation. It does so in part by reducing mortality and selecting for the life history changes that make advanced social learning possible. But it also makes collective scavenging and access to predators' kill sites possible. It may also preadapt hominins psychologically and socially to the collective suppression of dominant, expropriating males, thus opening the door to collective foraging, central-place foraging, and (more generally) variance-reducing strategies. Many such strategies rely on trust and secure possession of one's resources.

Many accounts of hominin evolution are structured around identifying a critical breakthrough, one that explains the unique features of human life. Hrdy and Wrangham exemplify this approach to our evolution:

> Without a doubt, highly complex coevoluationary processes were involved in the evolution of extended life spans, prolonged childhoods and bigger brains. What I want to stress here, however, is that cooperative breeding was the *pre-existing condition* that permitted the evolution of these traits in the hominin line. Creatures may not need big brains to evolve cooperative breeding, but hominins needed shared care and provisioning to evolve big brains. Cooperative breeding had to come first. (Hrdy 2009, 277)

> Once they kept fire alive at night, a group of habilines in a particular place occasionally dropped food morsels by accident, ate them after they had been heated, and learned that they tasted better. Repeating their habit, this group would have swiftly evolved into the first Homo erectus. The newly delicious cooked diet led to their evolving smaller guts, bigger brains, and reduced body hair; more running; more hunting; longer lives; calmer temperaments; and a new emphasis on bonding between females and males. The softness of their cooked plant foods selected for smaller teeth, the protection fire at night enabled them to sleep on the ground and lose their climbing ability, and females began cooking for males, whose time was increasingly free to search for more meat and honey. While other habilines elsewhere in Africa continued for several hundred thousand years to eat their food raw, one lucky group became Homo erectus—and humanity began. (Wrangham 2009, 193–194)

My approach is quite different. There is no key adaptation or magic moment. Rather, the elaboration of human cooperation depends on the evolution and stabilization of a set of positive feedback loops that connect technology and the division of labor with cooperation, with social learning, and with informationally engineered developmental environments. Here as elsewhere, we are creatures of feedback. My aim in this chapter is to chart this feedback; in particular, to identify the contribution of reproductive cooperation to cooperation more generally.

## 4.2 A Cooperation Complex

As I have just noted, one of the persistent themes of this book is to resist key-innovation models of human evolution and instead to picture our evolution in terms of positive feedback links. We evolved via coevoluationary interactions among a suite of cognitive, behavioral, and social capacities. One of these is our capacity to cooperate. In chapter 1, I argued that cooperation is itself a suite of coevolving components, emphasizing the intimate connection between information pooling and cooperative foraging. Ecological cooperation selects for information sharing, and information sharing makes cooperative foraging more profitable and less risky. The critical premise of my argument was that cooperative foraging (of the hominin variety) depends on technology and expertise and hence selects for information sharing at and across generations. Cooperative foraging generates great profit by taking advantage of information sharing and thus generates surpluses that make the development of better technology and more extensive expertise possible. The aim of this chapter is to further develop this model, but also to adjust it in the light of an alternative perspective. That alternative perspective emphasizes the role of reproductive cooperation in human evolution. I incorporate a modest version of that alternative but reject a more extreme one. I begin this section by briefly rehearsing key features of information sharing and cooperative foraging before turning in some detail to reproductive cooperation.

*Information pooling revisited.* Human social life depends on information pooling and social learning. We are obligate, habitual, inveterate, and adapted social information pumps, sucking information and expertise from our social partners. Moreover, human children have long childhoods: they have plenty of time to absorb the accumulated information and skills of their parents' generation before their time budget is blown by having to earn their own living. This seems especially true of forager cultures, with their tolerance of children's play and exploration. Subsistence farmers tend to use their children's labor more intensely, and from quite early ages (Hewlett et al. 2011). Human life depends on this informational commons and has done so at least since the regular harvesting of large game and the regular use of fire, and this may be as deep in time as two million years. Ever since those lifeways were established, we have depended on technology and technique to extract resources from our environment. Yet foragers do not themselves invent the technology, technique, and lore they need to extract resources from a recalcitrant world. They may improve on the informational resources that they inherit, or fine-tune those resources for their specific circumstances. But they inherit essential cognitive capital.

In the "embodied capital" model of Robson, Hill, Gurven, and Kaplan, this flow of information across the generations is so important that human life history has shifted to accommodate it by extending childhood and adding adolescence (see, e.g., Gurven 2004; Hill and Kaplan 1999; Kaplan et al. 2005; Kaplan et al. 2000; Robson and Kaplan 2003). This suggestion remains controversial. No one doubts that foraging is a skilled activity or that young foragers learn a good deal from their parental generation. But perhaps children are so well equipped for fast, accurate, high-volume social learning that they complete learning before they complete growing. Still, even if it does not take twenty years to acquire foraging expertise, that expertise depends on extensive social learning.

*Ecological cooperation.* Information sharing is coupled with economic and ecological cooperation. Obviously, economic life in all mass societies depends on collective action: in our social world, many essential tasks cannot be completed by a single individual acting alone. Modern life depends on collective action, coordination, and division of labor. But persuasive evidence demonstrates that collective economic action and division of labor are ancient features of hominin lifeways. Meat consumption became a key part of hominin diets, probably via increasingly aggressive scavenging as a precursor to actual hunting, and it did so long before any evidence of high-velocity projectile weapons. If the impressive Pleistocene predators were dispossessed of their kills by volleys of thrown hand axes and stone-tipped javelins, we can safely assume that this was not a solitary (or even a small-group) activity. Humans evolved a distinctive foraging style: it was information rich, cooperative, and tool assisted, and it targeted high-value resources (Sterelny 2007). Both *sapiens* and Neanderthals depended on animal resources that could only have been captured by bands acting cooperatively, for both species have been successfully hunting large game for hundreds of thousands of years. Indeed, collective foraging may date back at least to the last common ancestor of the two most recent hominin species. Early in Martin Jones's *Feast*, we are treated to a vivid reconstruction of collective consumption at a horse kill site at Boxgrove, England, dated to about 500,000 years ago (Jones 2007). Without the ability to kill at a distance, groups can take large game only by acting together.

*Reproductive cooperation.* Humans cooperate reproductively, not just economically. Children depend on their adult protectors, often for many years. Though children forage in many traditional communities, in very few do young adolescents gather all the resources they need. There are usually periods in the year when resources are hard to find. Even when children and young adolescents forage profitably, they typically use technology that adults provide. Younger children, of course, are even more likely to need adult resources and protection. Children throw a long resource shadow over their mother's future; Sarah Hrdy estimates that it takes thirteen million calories to raise children from birth to full independence (Hrdy 2005, 15–16). Yet human mothers wean their children early, on average between two and three. Chimps wean at about five years, orangutans even later (Kennedy 2005).

One difference between humans and chimps is that human fathers invest in their children, providing some mix of protection, resources, and direct care. Older siblings also often contribute to the family economy, again through some mix of provisioning and direct care of young siblings. But according to Kristin Hawkes, Nick Blurton Jones, and Frank O'Connell, the most distinctive element in reproductive cooperation is the role of grandmothers (Hawkes 1994, 2003; Hawkes et al. 1998; O'Connell, Hawkes, and Blurton Jones 1999). Menopause is an unusual feature of human life history: many women live for many years as active and competent agents after the birth of their final child, and these active grandmothers often play an important role in the care of their grandchildren. Chimp populations do not have this demographic profile. Chimps do not live as long as people, and female chimps do not have long and active postreproductive lives. Hawkes, Blurton Jones, and O'Connell argue that this feature of human demography is not just distinctive but crucial. Reproductive cooperation was the foundational form of human cooperation, and grandmothering was the critical adaptive breakthrough in the evolution of the distinctive form of human reproductive cooperation. In its boldest form, this idea challenges the basic assumptions of sections 1.3, 1.4, and 2.1: that human evolution was structured by the coevolution of cross-generational social learning with a new, collective mode of extracting resources from the environment. The defenders of the Grandmother Hypothesis are skeptical about collective foraging and about the idea that hunter-gatherer life depends on deep, extensive social learning.

The idea that humans are cooperative breeders need not imply that cooperation occurs only between mothers and their adult daughters. Both Sarah Hrdy and Karen Kramer point out that mothers are supported from many sources (Hrdy 2009; Kramer 2010); indeed, not always by kin. Older siblings, fathers, and the mother's own siblings can all be important. The Grandmother Hypothesis itself, as we shall see, emphasizes the role of *provisioning*: of supporting the mother by providing her weaned children with food. But in fact other agents support mothers in many ways: by protecting and supervising children through babysitting (easing the mother's time budget); by carrying infants, which is important in mobile forager cultures; by providing emotional and informational support to mothers; and by directly provisioning them. Provisioning includes breast milk in many forager cultures, and that is not a service grandmothers can provide.

However, while support undoubtedly comes from many sources, Hrdy shows that when she is still alive and kicking, a mother's mother often does play a special role. Grandmothers are *committed* to grandchildren. They no longer have direct reproductive interests that compete with allomothering, and when they must choose, they prioritize the youngest and most vulnerable, where their aid makes the biggest difference. However, mothers cannot rely on having a living, competent mother to offer aid. Grandmother survival rates have many uncertainties, but Hrdy's best guess is that in quasi-contemporary forager cultures, women have somewhere between a 25 percent and a 50 percent chance of having a surviving mother. Alloparents might be essential to successful human reproduction, but grandmothers could not be. So one way I will rework the Grandmother Hypothesis is by emphasizing the role of other sources of care in reproductive cooperation. More importantly, though, I resist the idea that reproductive cooperation played a foundational role in the evolution of cooperation. I think instead that reproductive cooperation was one leg of a cooperation tripod.

For Hawkes, Blurton Jones, and O'Connell, the pivotal member of the hominin lineage is *Homo erectus*, for while the evolutionary changes that resulted in modern humans took place gradually and over a long period, one can see *erectus* evolution as the time when hominins became human. *Erectus* was the first of the out-of-Africa hominins,[1] the first species in our lineage to expand geographically and ecologically. Its anatomy became human: *erectus* hominins were our size, fully bipedal, and large brained (though not as large as ours). *Erectus* was sexually dimorphic to a similar degree. Perhaps they also evolved toward *sapiens* life history patterns, with a lengthened life span and a longer period of juvenile dependence. Hawkes, Blurton Jones, and O'Connell take these life history patterns as their point of departure. In comparison with chimps (and our presumptive common ancestor), we are longer lived, larger, and our children are dependent much longer. Until recently, the standard explanation of human life history patterns resembled the cooperative foraging model of section 1.3. A hunting hypothesis related the *erectus* geographic and ecological expansion to the evolution of collaborative hunting and central-place foraging (Washburn and Lancaster 1968; for a more recent version of the idea, see Stanford 1999). Humans consumed the proceeds of hunting in groups in domestic circumstances not dissimilar to those experienced by recent foraging cultures. The cognitive demands on this ecological style selected for lengthened juvenile dependence, to give children the time to acquire the information and skills they would need as adults. In contrast, the defenders of the Grandmother Hypothesis see hominin life history changes as largely a side effect of increased body size. At the same time, this revisionary model developed a skeptical reanalysis of the role of hunting in human evolution. Hunting, the skeptics argue, is male display rather than family economics, mating effort rather than breeding effort.

## 4.3 The Grandmother Hypothesis

Humans are large, long-lived mammals with long childhoods. The defenders of the Grandmother Hypothesis think that we have long childhoods because we are large, not because we need twenty years of education in foraging life. Among animals that breed slowly, producing a small number of offspring that must make their way in a crowded world, selection will favor larger body size and hence delayed sexual maturity.[2] Size and strength are important factors in many competitive interactions, so larger animals typically outcompete smaller ones. But nothing is free, and size has to be paid for by delaying maturity. Yet delay is inherently risky. If the parents are at high risk of dying before their offspring are independent, or if the offspring are themselves vulnerable, the risk is too great. For many animals, the adult survival rate is too low to allow them to risk long juvenile periods and a slow growth to adulthood. They are too likely to die before they recoup their investment in growth. But if extrinsic mortality rates decline and risk falls, animals can invest in size, delaying maturity. If the risk of sudden death is lowered, it is also worth investing in antiaging physiology, for the rate at which animals decay through aging is itself variable and responds to metabolic investment. So we should expect to see positive correlations between long life spans, large body size, and delayed sexual maturity.

Hominins fit this basic pattern: we are primates with long life spans, delayed maturity, and large body sizes. However, we contrast with other large-bodied, long-lived primates (and mammals) in that females often have long, active, and healthy postreproductive lives. Women often abandon direct reproductive effort long before they die. Why? According to the Grandmother Hypothesis, this life stage is an evolved response of the *erectus* lineage to the decreased risk of adult mortality in increasingly seasonal environments. The reduction in mortality indirectly extended time as a juvenile via selection for increased adult size and thus delayed maturation. Children stayed children longer. Yet they needed more care, for in dry seasons, children could not gather their own resources. As intensified climatic variation increased the seasonality of *erectus* environments, the hominin lineage faced a choice: either find dry-season resources and a way of delivering them to the young, or retreat to less seasonal refuges.

Washburn's hunting hypothesis was one candidate solution to this dilemma: meat and marrow are available in the dry season, and adult males (especially) deliver these resources to women and children in (extended) family units (Washburn and Lancaster 1968). Hawkes, Blurton Jones, and O'Connell have another suggestion: underground storage organs are the dry-season resource, and in fortunate families, an active grandmother finds and delivers these resources to the older children of her daughters. Grandmothers provision their weaned grandchildren, allowing their daughters to care for their infants, thus reducing the interbirth interval of families that have active grandmothers. Selection would thus favor a late-in-life switch of female strategy from direct reproductive effort to indirect effort, for the prospects of raising a late-life child would be poor. Selection would also favor investment in antiaging physiology. Modest periods as an aiding grandmother would be visible to selection, so active grandmaternal life could evolve (once extrinsic risk declined) *by degrees*. To improve her direct fitness, a potential grandmother going it alone as a late-in-life mother would have to survive as a competent agent through the whole period of the dependence of her final child. To improve her indirect fitness, she might need only to take care of one three-year-old through one dry season, thus enabling her daughter to squeeze an extra infant into her lifetime fertility schedule.

In short, the grandmother model has five key components. (i) It requires a reduction of extrinsic mortality, making delayed maturity profitable and opening the door to profitable investment in slowing senescence. (ii) That reduction in mortality took place in environments that were increasingly seasonal, hence weanlings could not support themselves year-round by their own foraging. (iii) In dry-season environments, late-mature women could generate a nutritional surplus; they could feed both themselves and a daughter's dependent child. (iv) Hominin populations contained a reasonable proportion of older adults. Selection in favor of abandoning late-life direct-breeding efforts in favor of grandparenting would not change hominin phenotypes if those hominins rarely survived to late maturity. (v) The social organization did not involve males staying in their natal group and females leaving.[3] For the model to work, mothers and daughters must live in the same group, with grandmothers able to recognize their grandchildren. In some breeding systems (probably including those of chimpanzees), grandmothers cannot identify their sons' children. Thus the mother–adult daughter relationship plays a special role in this model.

Sarah Hrdy shows, I think, that the residential assumption is reasonable. Women in deep time had access to their female kin. While male great apes typically stay in their birth groups, with females leaving (thus great apes are "patrilocal"), they show ecological and social flexibility. Females do not always leave their natal group. And despite reports to the contrary, forager societies are not predominantly patrilocal. There is often considerable variation between individuals in the one group, and in individual residence over time. For example, grandmothers will often move in with daughters when they are having their first child, even in patrilocal worlds (Hrdy 2005, 2009).

Hawkes, Blurton Jones, and O'Connell themselves are most concerned to defend the demographic assumption. They argue that in ancestral populations, it was not unusual for a mother to survive to be a grandmother. The literature often suggests that long life expectancies are the privilege of modern populations; the idea is that humans in traditional societies rarely lived to be old.[4] Not so, according to Hawkes, Blurton Jones, and O'Connell. They argue that long human life spans have a deep history, and foraging populations include older adults in significant numbers. Indeed, even the samples cited by Gail Kennedy, a leading skeptic, include older adults (Kennedy 2003, 2005).[5] Hawkes and her colleagues suggest that the basic pattern of *sapiens* seems to be essentially stable over the transition from foraging to agriculture to urbanization despite real changes in fertility and life expectancy. Preindustrial populations contain a significant fraction of postreproductive adults, even when average life expectancy is in the thirties or early forties. Such averages are compressed by large numbers of early deaths. Thus while we lack much direct evidence of the age structure of pre-*sapiens* populations, the demographic foundations of the model are credible, and helpful grandparenting will be visible to selection.

In summary, the idea that reproductive cooperation is a central component of human cooperation is plausible. In many cultures, mothers have shorter interbirth intervals than chimps, though humans are subadults for much longer, and weaned human children, but not weaned chimps, need not just food but a host of other resources too. These include shelter, clothes, protection, and social support. *Erectus* children needed many of these extra resources too. High-cost children together with shortened interbirth intervals signal reproductive cooperation. A modest version of the Grandmother Hypothesis is plausible. Grandparent-based reproductive cooperation was one part of the cooperation revolution in human life. Moreover reproductive cooperation positively interacts with other forms of cooperation. For example, it makes social learning more robust by giving young children access to more sources of information (Burkart et al. 2009). But I am skeptical of the more ambitious idea that the evolution of the nurturing grandmother was the key innovation—the adaptive breakthrough that drove hominin transformation. So I aim to incorporate reproductive cooperation into the cooperative foraging model of section 1.3.

I begin with the idea that grandmothering was only one element of reproductive cooperation. I then argue (perhaps more challengingly) that reproductive cooperation could not evolve *first*: its evolution depends on ecological and informational cooperation. Having argued for a coevolutionary perspective on reproductive cooperation, I then sketch positive interactions between reproductive, ecological, and informational cooperation. Finally, I turn to the issue of stability, important because many potentially profitable forms of cooperation are not evolutionarily stable.

*Grandmothers and other alloparents.* Grandmaternal aid was probably only one element of reproductive cooperation. Older siblings, aunts, and even fathers and paternal relatives are often important in *sapiens* social worlds (Kramer 2010). We have no reason to suppose that these alternative channels of alloparental support have more recent origins than grandmothering. The extension of subadult life made older siblings increasingly available as potential sources of protection, education, and support. Moreover, siblings are as closely related to one another as mothers are to their children, so an aging female who abandons direct reproduction can gather as much indirect fitness benefit as an aunt as she can as a grandmother. Aunting is as likely as grandmothering. And in ancient hominin social worlds, many females would lack a living mother. Even in optimistic assessments of survival, only about 50 percent of mothers would still have their own mother alive and active. On other estimates, the number falls to about 25 percent (Hrdy 2005, 16). If mothers needed support, much of it must have come from other sources.

Finally, what of fathers? Hawkes, Blurton Jones, and O'Connell are skeptical about the importance of fathers as allomothers, because they are skeptical about the idea that hunting is a provisioning tool. I dissent from their view of hunting in some detail in sections 4.4 and 5.5. But let me begin with a telling point from Gail Kennedy. She points out that growing children need specific resources. Weaned young children need more than calories; they need protein to fuel their postbirth brain growth (Kennedy 2005). Hawkes, Blurton Jones, and O'Connell identify underground storage organs as the key resource provided by grandmothers. But scavenged brains and bone marrow are the most plausible source of protein, and such scavenging is unlikely to be a grandmaternal specialty.

The early age of human weaning has serious costs: weanlings have a higher exposure to pathogens at an age when their immune system is less well developed, and when they lose the indirect immunological benefits of mothers' milk. The Grandmother Hypothesis assigns the benefits of early weaning to mothers. Early weaning, supported by allomothering, reduces the interbirth interval and increases lifetime fertility. Kennedy thinks the compensating benefits fall to the child. Earlier weaning meets the child's need to fuel postbirth brain growth, given the unsuitability of maternal milk to fuel that growth once children are roughly of weaning age. Without early weaning, a brain-growth-driven protein gap would open out as protein needs increase beyond the point that they can be sustained by maternal milk (Kennedy 2005). If this nutritional argument is sound, we may well be able to draw an important link between the extended childhoods and the habitual use of stone tools. The exploitation of large carcasses through scavenging and bone breaking (very likely in the face of competition from other scavengers and predators) gave access to the foods needed for childhood brain growth. So we come to a first suggested modification of the Grandmother Hypothesis: think of grandmothering as just one element of the evolution of reproductive cooperation.

*The preconditions of reproductive cooperation.* More importantly, it is a mistake to think of reproductive cooperation as the precondition of other aspects of cooperation. The three elements of the cooperation syndrome must have evolved together, for grandmothering could establish only in a social environment that was already cooperative. First, the provisioning pattern of grandmother support presupposes the presence of other forms of cooperation. Second, the life history model depends on reduced mortality, and ecological cooperation is the most plausible mechanism through which mortality might fall. Third, the harvest and use of underground storage organs depend on informational cooperation. Finally, the psychological capacities presupposed by grandmother provisioning were not available to a great ape with an enlarged brain at the beginning of the hominin cooperation explosion.

First, then, the grandmothering model tacitly assumes that the social environment is benign. Most underground storage organs cannot be consumed on the spot; they need to be processed. Moreover, they are often quite sizable packages of food. In a Hobbesian social world, grandmothers and infants with such resources would often lose them, and not just to occasional unfortunate encounters with the largest males. They would be vulnerable to junior and subadult males as well. In such social environments, the vulnerable need to adopt a feed-as-you-go foraging style, and harvesting and processing underground storage organs would not be viable. The social environment must have been at worst tolerant and unaggressive. In fact, though, the grandmother model presupposes more. Without ecological cooperation, erectines would have been vulnerable to the impressive guild of Pleistocene predators. Hawkes and her allies must suppose that grandmothers—alone or perhaps with a few other older women—were out, searching through the African bush, in the hungry season, when their help is most necessary. They were armed with nothing more lethal than digging sticks. This would have been a risky business—too risky, unless hyenas, lions, and other predators had already learned to be wary of humans, or had been reduced in number through competition and persecution, or both.

Second, according to the Grandmother Hypothesis, human childhoods expand through selection for large adult size, and this in turn depends on reductions in extrinsic mortality. What explained improved adult survival? It was not the result of a more benign external environment. To the contrary, Rick Potts (1996) has argued that the frequency and magnitude of environmental fluctuations increased in the second half of the human career; Peter Richerson and Robert Boyd have made the same point more recently, using recent data (Richerson, forthcoming; Richerson and Boyd 2002). While Hawkes, Blurton Jones, and O'Connell are silent on the reasons for the fall in mortality, the most plausible suggestion is that increasing cooperation reduced risk by defending more effectively against predation and by dampening variance in food supply. Predation is not just a problem for lone grandmothers. As Wrangham (2009) points out, habilines and erectines were not well suited to nesting at night in trees, as chimps do. Their morphology required them to sleep on the ground. With many predators active at night, this too would be dangerous, unless they had some system of collective defense, perhaps involving controlled fire, as Wrangham suggests. Social support during times of illness and injury might also have been important. Rudimentary nursing—providing food, water, and protection— greatly improves the survivability of serious injury and illness.

Third, underground storage organs—supposedly the key grandmother resource—are rich. But finding them can be challenging, and more crucially they often require significant processing. Many are inedible without cooking, or washing and leaching to remove toxins. Often the techniques involve complex cultural adaptations; the processing technique is not a one-step innovation that a lucky individual could stumble on. Yet building complex adaptations is a signature of information pooling and social learning, one of the crucial elements of the cooperation syndrome described in section 4.2.

Finally, provisioning by grandmothers depends on minds already adapted to cooperation and sharing, for it depends on a psychology able to resist temptation. Chimps find food—actual, visible, right-here-and-now food—a mesmerizing, irresistible temptation. Recall, for example, Sarah Boysen's wonderfully amusing experiments in which chimps had to point to the smaller of two piles of food to get the larger one, a task they found impossible (Boysen 1996). A form of cooperation that requires an agent to spend real effort searching for food, finding it, extracting it, processing it, and then freely giving it to another agent cannot be a *base case* of cooperation, a form of cooperation from which much else evolved. Such cooperation demands highly developed impulse control and the capacity to stay on track with a complex, temporally extended set of actions, without any immediate reward. The emotions of affiliation alone will not deliver such a capacity: the history of lustful encounters at parties shows that the emotions of affiliation do not give you for free the capacity to manage temptation. Such capacities are the result of a coevolutionary trajectory which amplifies cooperation and cognition together. They are not available at the starting point of such a trajectory.

In short, grandmothering evolved in a milieu of other forms of cooperation, both reproductive, informational, and ecological. It was part of an evolving cooperation suite, perhaps not even an especially central element in that suite. Grandmothering might explain menopause, but menopause does not explain the cooperation explosion that transformed hominin life.

However, while reproductive cooperation is not the key innovation from which the distinctive forms of human cognition and cooperation evolve, it is a crucial ingredient in the mix. The hominin move to shared care had profound consequences for infant development. The developmental environment was richer and more varied, involving interaction with males, other juveniles, and especially other adult females. Sarah Hrdy (2009) points out that in many primate worlds, infanticide is a real threat, so mothers are extremely vigilant and possessive, and infant primates often interact only with their mother. If extensive sources of alternative care are available, the young gain social experience from many agents and have more sources of information open to them. There are more models, more redundancy in the social signal. In sum, the social (and probably environmental) inputs were much richer.

The selective environment changes, too, as care comes from more sources. Even if an infant's erectine mother was unconditionally attached and committed, the commitment and care from her affiliates was up for negotiation. An infant's own actions can influence for better or worse the extra care available. So a world of reproductive cooperation will select for infant appeal. As Hrdy (1999, 2009) notes, the very young are often sensory traps for adults. But an alloparenting world selects for more than cuteness. It will select for (a) infant monitoring of mothers and others, (b) infant awareness of the differences among the others, and (c) infant awareness of others' responses to its own action: awareness of joint attention and action. Once care is a negotiable quantity, babies are under selection for social skills, and that might help explain the recent results in developmental psychology suggesting that infants have a surprisingly rich theory of mind (Baillargeon, Scott, and He 2010).

*Stability.*   Finally, if reproductive cooperation is an early form of cooperation, we must show that it would be stable before the invention of norms, policing, and investment in reputation. These tools stabilize cooperation, but they are invented only in worlds in which cooperation has long been important. One advantage of the grandmother model as a model of reproductive cooperation is that it raises no problem of defection. The grandmother's and the daughter's fitness interests are aligned. Grandmothers increase their inclusive fitness, for they boost the survival prospects of their grandchildren with little opportunity cost. An aging mother is unlikely to survive long enough to see a child of her own through the prepuberty crisis, so she loses little in concentrating her efforts on grandchildren.

However, grandparents are not the only alloparents. What of others? Adolescent girls are often important sources of help. Arguably they gain important information by helping to care. Nurturing requires a set of learned skills. It is not innate, probably because what a mother needs to do is sensitive to specific circumstances. There is much to learn, and practice helps. In many mammal species, first-time mothers have below-average breeding success (Hrdy 2009). In addition to the benefits of upskilling, older siblings gain some inclusive fitness benefits, especially when they care by providing babysitting and protection. Such care is important but cheap.

Help from adults of the mother's generation is perhaps more subject to cheating problems. When help comes from the mother's siblings, some modest inclusive fitness benefits will accrue. But reciprocation and indirect reciprocation are probably important in explaining the exchange of care between adults. Hrdy documents the centrality to fitness of social networks of support in forager worlds, and these networks are maintained and reinforced by aid (see, e.g., Hrdy 2009, 14–16). Giving aid to parents invests in the relationships, and shows commitment to them. These networks will often be crucial to the helper's own situation if and when she hits a bump in the road. I take up the problem of the stability of these networks of mutual support, given the problem of free riding, in the next chapter. But reproductive cooperation does not seem vulnerable to decay through free riding.

It can begin with kin support and subadults helping for practice. It might then be elaborated by becoming one important aspect of networks of affiliation and reciprocation. These networks are maintained by norms, gossip-reputational effects, and the capacity to recall and value contributions (and contribution failures). Moreover, much of the reciprocation-based alloparental cooperation involves small costs and large benefits: babysitting, vigilance, protection. The most likely sources of aid have little motive to cheat. This is high-value, low-risk cooperation, and for this reason it is plausibly seen as one of the foundations of the evolution of human cooperation. This may then select for cooperative dispositions that can be elaborated for broader cooperative networks in conjunction with newly evolved cognitive capacities and invented cultural tools.

## 4.4 Foragers: Ancient and Modern

In sections 1.4 and 2.1, I argued that human life histories have been structured by the demands on social learning. Adult life is delayed, and cross-generational resource flows organized, because high-expertise adult foraging is highly profitable. Adult foragers, especially males, deliver a lot of resources. But the expertise necessary to generate this profit takes many years to acquire. Childhood is an adaptation to the extensive demands of social learning; it can be afforded only because that learning makes adult activity far more profitable than it would otherwise be. This view of human life history takes hunting to be centrally an *economic activity*. It is provisioning rather than signaling. The costs of learning are worth paying because adult provisioning, especially adult male provisioning, is much more profitable as a result of this investment. We thus link models of the sexual division of labor with those of human life history. Those who think of hunting as primarily economic think of lengthened juvenile periods as adaptations for the accumulation of expertise. Those skeptical of this view of hunting think of long childhood as a side effect of the extension of adult life and of the selection for increased adult body size.

In thinking about ancient foragers, modern models loom large (O'Connell 2006). They are a lens through which the remains of ancient lives have been interpreted. This poses a methodological problem, for ancient and modern foragers are profoundly different. I argue that despite these differences, we can use modern data to answer important questions about ancient lives. In favorable cases, we can correct for the differences between ancient and modern foraging worlds. In particular, in considering the cooperative foraging model of human life history, modern data offer a *conservative test*. The ancient-to-modern transition would tend to damp down a class of important features of ancient forager lifeways, ones that make cooperation more important. So if we still find those features playing a role in the lives of modern foragers, we can reasonably project them back onto the lives of ancient foragers.

Foragers and their world have been transformed over the last 150,000 years (for a good review, see Stiner 2002; for a review of foraging changes over the whole hominin trajectory, see Foley and Gamble 2009). One important transformation has been in the environment of cooperation. Maintaining stable cooperation has become more problematic. In sections 1.3 and 1.4, I argued that humans extract resources from their environment in a unique way: via collaborative application of expertise. Foraging is both cooperative and dependent on expertise and technology (Sterelny 2007). Over the last 40,000 years, the pattern of collaborative foraging has changed markedly. The so-called broad-spectrum revolution involved, inter alia, a shift away from large and medium-size game (ancient *sapiens* specialized in ungulates) to small and medium-size game, and to marine and other resources (Stiner 2001). The broad-spectrum revolution selected for specialization (as different techniques and technology are needed for hunting, say, rabbits rather than wildfowl or fish). It also selected for much more small-group and individual hunting. At the same time, this shift reduced both variance in daily success and the size of individual items taken. Kills were much more frequent, but much less was provided per kill. The distinction between hunting and gathering became less marked. So in many environments, hunting became less collaborative, and the need for reciprocal sharing less pressing.

This social effect of the broad-spectrum revolution was intensified by the projectile revolution. Until the Middle Stone Age, perhaps the late Middle Stone Age (within the last hundred thousand years), toolkits were simple, and weapons fairly short range. A thrown, stone-tipped spear of the kind used to kill horses at the Boxgrove site 400,000 years ago is unlikely to have a kill zone much beyond ten meters.[6] If humans between 200,000 and 400,000 years ago had regular access to the meat of large animals, either by direct kills or by expropriating the kills of other large carnivores, technique and cooperation must have been crucial. There seems to be no evidence even of thrown javelin-style projectile weapons until the last hundred thousand years. John Shea (2009) argues that finer, narrower points of the kind suitable for mounting on thrown javelins began to appear in Africa in the late Middle Stone Age. His views are based on a detailed analysis of stone points, comparing Middle Stone Age points to points from javelins and stabbing spears in ethnographic collections. This analysis will not tell us whether Middle Stone Age javelins were arm powered, or whether Middle Stone Age Africans had spear-throwers. But according to Marlowe (2005), between perhaps 30,000 and 20,000 years ago, humans added spear throwers, the bow and arrow, and poison darts to their arsenal.

All of this matters because the ability to kill at a distance changes the environment of cooperation. It became possible for individuals or small groups to kill large animals in relative safety. Large groups that hunt and kill together can share on the spot. The profit of joint activity is accrued together and in full view of all, so no informational problems arise in policing cooperation. Identifying and agreeing to a fair division of a joint resource is much less problematic if everyone is a roughly equal partner in a joint activity (Ostrom 1998). Division becomes more problematic once individual success becomes highly variable (as individuals hunt alone or with favored partners), once the range of resources expands (making commensurability an issue), once role specialization becomes important, once reciprocation extends over time, and once individuals spend much of their time, and enjoy much of their success and failure, away from the eyes of the many. Cooperation is most stable in small, homogeneous groups.

We have little robust evidence on group size over the last few hundred thousand years, though it is usually supposed to have increased (Dunbar 2003). We have more robust evidence of increasing heterogeneity and role differentiation, exacerbating the cognitive challenge of monitoring fairness. In sum, then, in comparison to ancient forager lives, cooperation in modern forager life is probably more fragile because it is more optional. Indeed, I conjecture that the evolution of explicit and articulated norms is in part a sign of stress on cooperative defaults. Whether that is right or not, the main point is that if contemporary and near-contemporary forager lives are typically cooperative and egalitarian (Boehm 1999, 2000; Gurven 2004; Hewlett et al. 2001), we should expect ancient forager lives to have been even more cooperative and egalitarian.

In their recent review of the evolution of social organization, Kaplan, Hooper, and Gurven (2009) argue that egalitarian, cooperative social organization depends on four factors: (i) key resources cannot be monopolized by one or a few agents; (ii) all adult economic activity is highly skilled; (iii) female and male roles are complementary, and both are important; and (iv) communities are small, so that bottom-up mechanisms of norm enforcement, coordination, and decision work. Forager communities are still small, but some of these factors have been somewhat eroded in recent forager worlds. But they were likely to be in full play in ancient forager communities.

*Resource monopolies.* Some forager cultures (so-called complex foraging cultures) are organized around minority control of rich resources. But Shultziner and his coauthors (2010) argue that rich, defendable resources (like the salmon runs of the Pacific Northwest) emerged only in the Holocene. The Pleistocene climate was too unstable (and often too harsh) for there to be rich and predictable resource hot spots that a few agents could seize. While this is probably an overstatement (seal rookeries are rich, predictable, and defensible), the contrast between the unpredictable Pleistocene and the more stable Holocene is real and important (Richerson, Boyd, and Bettinger 2001).

*Skill.* The cognitive demands on a forager's life have also changed. In some respects, those may well have increased through the broad-spectrum revolution and the expansion of material culture. Instead of having to develop a detailed understanding of a few key target species, broad-spectrum foragers had to have a feel for the natural history of many. Instead of having to master the techniques of producing a limited toolkit, they had to master the production of a much more varied technology. In most respects, though, the cognitive demands are less onerous (especially if the production of physical symbols allowed them to store information in the world). The use of dogs reduces the pressure on tracking skills. The use of projectile technology makes the skills of concealment and stalking less essential.[7] Most recently, trade with the nonforaging world has reduced the demands on artisan skills.

It is hard to estimate the informational demands on preindustrial foragers from modern data. Researchers have tried to estimate these demands, to determine whether the link between foraging success and age depends on size and strength or twenty or more years of social learning (Bliege Bird and Bird 2002; Blurton Jones and Marlowe 2002). But as foraging peoples have now supplemented or replaced home-built technology with store-bought equipment, empirical tests focus on foraging itself, rather than on the cognitive demands of being equipped to forage. With few exceptions, they measure only one aspect of the skill set of ancient foragers. Moreover, observational methods of measuring success in the wild treat individual success as a reflection of individual competence (Bock 2005; Gurven, Kaplan, and Gutierrez 2006; MacDonald 2007). But individual foraging activity may well be influenced by consultation and advice from others. An egalitarian social world influences individual production, not just distribution. In short, our direct experimental information on foraging skill is meager, ambiguous, and narrowly based. Even so, from what we do know, these skills remain extremely impressive, as Louis Liebenberg (1990, 2008) demonstrates in his documentation of the skills required for tracking game in southern Africa.

*Complementarity.* Finally, the economics of hunting have changed, dampening down the differences between male and female activities. Hunting that targets small game is more like gathering. Small-game hunting has less variance, is less physically risky, and is more of an individual activity. The take is less communally shared, and each hour of effort is likely to show a lower rate of return. The broad-spectrum revolution was probably driven by foragers being forced to seek less-rewarding targets as encounter rates with the more desirable targets fell. With the massive depletion in most habitats of medium-size to large herbivores, hunting by contemporary and near-contemporary foragers is probably much less profitable. They live in a more Malthusian environment: the human footprint on the local resource profile has been heavy, persistent, and depleting. Hunting targets are less abundant, and those that remain are targeted by more.

The gap between male and female returns may have closed not just because male hunting has become less profitable as prime target species declined, but also because foraging by both women and children has become more valuable. Although children are rarely self-sufficient, in favorable circumstances, they can contribute significantly to the family economy. That may be a relatively recent development. In many environments, children's semi-independent foraging is possible only because predators have been largely eliminated, and those that survive have learned to avoid humans. This is probably a relatively recent, Holocene development, one that has eased the energetic demands of children on their parents. As noted in the discussion of *erectus* grandmothers, to a lesser extent, the same may be true of female gathering. If we set aside the threat from other humans (an issue I take up in chapter 8), independent foraging by women and older children is surely much safer now than it was 100,000 years ago, when we were just one of an impressive set of African Pleistocene predators.

Thus contemporary and ancient foragers did not face similar ecological, social, and sexual decisions with similar resources. Ancient hunting was probably more routinely cooperative, with large kills shared on the spot among quite large hunting parties. The most valuable prey was more abundant, and so the profit of hunting was probably higher. But the prey was more dangerous, especially in comparison to the available weaponry. Gathering by women and children may well have been constrained by predation threat. Ancient hunters had to manufacture all their own equipment and detect and close with their prey. They had to track prey with their own observational powers rather than piggyback on those of dogs. Yet the results of hunts are still regularly shared. Hunting is often more profitable to hunters and their family than gathering. Foraging in general and hunting in particular are highly skilled; size is not everything, and older and more experienced hunters sometimes do better than young adults in their physical prime (see, e.g., Walker et al. 2002).We can extrapolate these traits and more to ancient foragers. Middle Stone Age hunting was communal and cooperative, coevolving with, supporting, and enhancing reproductive and informational cooperation. But it also depended on those other forms of cooperation. No doubt hunts generated socially important information. Hunters cannot help but reveal their character, knowledge, and skills, for better or worse. No doubt these informational effects came to influence the mode and social context of hunting. But hunting became established in the hominin lineage primarily as a mode of resource extraction—as an example of ecological cooperation—and so it remained. Or so I shall argue in the next section.

## 4.5  Hunting: Provisioning or Signaling?

Hawkes, Blurton Jones, and O'Connell do not think of hunting as primarily a provisioning activity; in particular, large-game hunting functions to send a costly signal: hunters advertise their quality. The crucial idea of costly signaling models is that agents have an interest in advertising their high quality, but less-fit agents have an interest in overstating their quality. Since quality is not directly observable, genuinely high-quality agents and their audience both have an interest in identifying signals that honestly advertise quality. Costly signals enable the audience to discriminate between high-quality agents and pretenders, because the signals are not just costly; they are differentially costly to signalers (Grafen 1990; Zahavi and Zahavi 1997). Frauds cannot afford the signal, and reliability is stabilized by *differential* cost (Saunders 2009). In competitive interactions, for example, it is in the interests of high-quality agents to signal their fighting power, and in the interests of others to read those signals. But since fighting power is not directly observable, it is also in the interests of weak individuals to deceive: to signal that they are powerful even though they are not. To be honest, signals of power must be costly, perhaps involving risky displays or high-energy activities. Though such signals are costly to all, they are not equally costly for all: for example, flirting with danger is less risky for agile, fit, strong agents. At the limit, only those who have the crucial characteristic can afford the cost of sending the signal, and thus the signal is an honest measure of quality.

Costly signals and their role in human evolution are a prime focus of the next chapter, and I discuss hunting in section 5.5. For now, my primary purpose is to argue that hunting should be seen largely, though not entirely, as an economic activity. More particularly, hunting established and became central to hominin life as an economic activity. That said, three caveats: First, some cases of forager hunting are, probably, primarily costly signals; however, I do not think these are typical. Second, I agree that hunting is not purely instrumental. It plays other roles in forager life, and I return to these in section 5.5. Third, even if hunts are not signals, they are cues. Hunting produces stress and fatigue as well as meat, and stressed, tired agents leak information about their character and capacities. In village-sized worlds in which reputation matters, others will read these cues. So even if hunting is a stable feature of foraging life through the resources it generates for hunters, their kin, and their allies, in reputation-driven social worlds, hunting is likely to play an informational role as well. Moreover, once hunting is important as a cue, we are likely to see secondary effects. Hunting, hunting lore, customs, and norms are likely to be influenced by the impacts of hunting on reputation.

Hawkes and her allies think that large-game hunting is centrally a signal or display. This is not an ancillary, additional facet. Rather, hunting, the idea goes, is an unfakable signal of quality: only the genuinely fit hunt successfully (Hawkes 1991; Hawkes and Bird 2002; Smith and Bliege Bird 2005). In their view, the primary returns to hunting are social: status, prestige, and sexual access rather than hunters provisioning their family directly and indirectly via reciprocation. Despite its popularity, I think this model is unpersuasive. I begin with an ambiguity in the model. It is not obvious why hunting is differentially costly for the less expert. If hunting was an individual activity, the differential costs would be time and energy: it would take the less-expert hunter longer to produce a given quantity of game. But even in recent forager cultures, much, perhaps most, hunting is a collective activity. When hunting is a collective activity, it is not obvious that the less expert pay higher costs per kill. The whole group might do less well because a few of its members are less skilled, but do the less expert do less well? Another possibility is that hunting is riskier for the less expert. Before the widespread use of projectile weapons, when hunting involved close encounters with large game, there probably was a difference in risk. But once hunters come to use woomera-assisted javelins, bows and arrows, or poison darts, they attack from a distance. Expertise is more likely to have an impact on encounter and kill rates than risk: experts are more likely to find targets and to hit them when found. Indeed, perhaps the most plausible "cost" is the reputational cost of participating ineffectively. But that raises the issue of why, if the display model is right, the less able hunt at all. On this, more shortly. My first reason for skepticism is thus that the cost differential is not well defined.

Second, the fact that much hunting is collective challenges the model. We should expect costly signaling to be an individualistic activity, as is, for example, building a new art gallery for a city. But this does not fit the pattern of hunting in many forager societies. Hunting is often collective and cooperative rather than individual (see, e.g., Alvard and Nolin 2002 on cooperative whale hunting). Indeed, as I argued in section 4.4, only the invention of penetrating projectile weapons made individual and small-group hunting of large game possible. Such projectile weapons have much more recent origins than regular hunting; foragers may first have used the bow as recently as 20,000 years ago (Marlowe 2005, 64). Dates are never certain, but unless current estimates are grossly wrong, humans hunted large game for hundreds of thousands of years without projectile weapons (except, perhaps, hand-thrown javelins). Large-game hunting played a central role in human economies long before humans could kill at safe distances in small groups. Hunting became central to human life as a collective, cooperative activity.

Of course, it is possible to show off in a collective activity, as showboating sports stars so amply demonstrate. But showing off is most evident when onlookers (or participants) can reliably identify individual contributions, and that usually involves role differentiation. In fast-moving, collective encounters, it is often difficult to identify a specific individual's effects. One indication of how difficult it is to identify such effects in a confused melee comes from the aerial warfare of World War II. On each side, combatants massively overclaimed their kills, despite the efforts of intelligence and evaluation officers to enforce conservative claims. Double and triple counting was epidemic. I do not want to exaggerate the problem of identifying the successful. Foragers who regularly hunt together certainly accurately assess one another's abilities. Even so, if the core rationale for hunting was to signal individual quality, we would expect it to be, like the Masai's lion hunts, an individual activity. Differences in individual capacity are exposed more widely, rapidly, and unambiguously in individual activity.[8] That is why universities standardly examine students through individual rather than collective work.

Third, we expect sending costly signals to be not just an individual activity; it should be a minority activity (see, e.g., Smith and Bliege Bird 2005). The costly signal is a display of *unusually high quality*. In the simplest Zahavian model of costly signals, only high-quality individuals can afford to send the costly signal that shows their quality. In this version of the costly signaling model, if hunting and sharing are the signal of quality, only high-quality males will hunt and share. Poor hunters should keep what they catch and scrounge what they can, or more likely they should abandon hunting altogether and forage like a girl. If hunting-and-sharing genuinely is individual signaling for individual advantage, dropouts need not fear being excluded from the distribution of product. They are not defecting from a cooperative activity; others should welcome their withdrawal from competitive signaling. Hunting should be like playing baseball in America: the best hunt, and the rest watch.

We can understand costly signaling in alternative ways. According to an alternative model in which *signal intensity* reflects male quality, we would expect high-quality males to hunt more frequently, or in more challenging circumstances. The signal is not hunting itself but hunting effort, and that effort should covary with male quality. One way or another, we should observe marked within-group differences in male attitudes to, and participation in, hunting, differences that line up with male quality. Neither expectation seems to fit the ethnographic record well. In many cultures, males are notoriously addicted to hunting. The centrality of hunting in the lives of less-expert males remains unexplained.[9]

Finally, while some forms of male hunting make little sense as family provisioning, and plenty of sense as signaling, these seem to be the exceptions rather than the rule. Costly signaling is a natural model of some forms of prosocial behavior in human life: feasts given by village leaders ("big men"), conspicuous gift giving to charity, and perhaps supererogatory displays of courage in intercommunal conflict (Smith, Bird, and Bird 2003). And there do seem to be examples of male hunting that fit the signaling paradigm. A persuasive example is that of Meriam turtle hunters, for not all the males hunt, and few are hunt leaders. Moreover, Smith, Bird, and Bird are able to show that turtle hunting does not make sense as an economic activity: there are alternative, and more rewarding, sources of protein (and, in the right season, even of turtle meat). Successful catches are shared, with no evidence of family bias or reciprocation. Finally, Smith and his coauthors are able to show that turtle hunters are fitter: they have a higher reproductive success, both because they acquire mates earlier than nonhunters and because they have higher-quality mates (Smith, Bird, and Bird 2003). While these authors cannot rule out the possibility that success at turtle hunting and superior fitness have a common cause, the idea that turtle hunting is a reliable signal of high quality is certainly plausible.[10]

Likewise, as Stephen Downes pointed out to me, the costly signaling model naturally fits hunting large predators, especially when the hunt is part of a recognized local ritual, as in Masai lion hunts (Ikandal and Packer 2008). Some contemporary hunts probably are signals in sexual competition. But though our information is patchy and often qualitative, the Meriam seem to emerge as the exception, not the rule. A persuasive reason for seeing Meriam turtle hunts as a display is that Meriam turtle hunters do not seem to share with their group in expectation of reciprocation, nor do they use their catch to preferentially provision their families. Gurven and Hill (2006) argue that this pattern of sharing is exceptional. They argue that hunters typically have considerable influence on the distribution of their hunting profits, and that these influences result in a bias toward kin (as a provisioning model would predict), and toward those that reciprocate. Food is given to those who give. Male sharing is either kin based or contingent. If Gurven and Hill turn out to be right that this is the general rule, then sharing large catches with nonkin is likely to be a variance-reduction strategy.[11]

Moreover, hunting does seem to be an efficient form of provisioning. Indeed, it is hard to reconcile the signaling model with the fact that in some high-Arctic cultures, plant food and small game are scarce, and families depend on male hunting. A signaling model would need to show that the targets of hunting are chosen for their difficulty, not their return rate. Moreover, while few studies are detailed and quantitative, several show that male hunting generates an average rate of return around twice that of gathering (Kaplan et al. 2005; Kaplan et al. 2000; Robson and Kaplan 2003). Furthermore, calorie counts understate this difference: meat is much richer in important macronutrients (proteins and fats) than vegetable-based foods. But even if on average, hunting is no more productive than gathering, specialization in different resources will generate synergies when the family unit has a variety of resource needs and when each resource can be gathered only by individuals with special skills and equipment. Finally, while it is true that returns for male hunting are more variable than those of gathering, variability does not undermine hunting's economic importance. Storage, sharing with other hunters, female gathering, and hunting small game all buffer variance. The transition from ancient to modern foraging is likely to have improved the fit of the signaling model with the data rather than eroded it. But even when we focus on contemporary foraging, the model fits the general trends poorly.

Time to sum up the main message of this chapter. Reproductive cooperation was indeed a key component of the cooperation complex, a complex whose evolution and elaboration transformed hominin social life and hominin minds. Grandmothers probably played an important role in the establishment and stabilization of reproductive cooperation. But the more important we believe reproductive cooperation to have been, the more mothers must have had other sources of support. In contrast to the most ambitious versions of the Grandmother Hypothesis, I have argued that male support was crucial. But males were able to support their families largely because they were embedded in networks of ecological and informational cooperation. These are as pivotal and as early as reproductive cooperation. In the next chapters, I turn to the stability of these networks of cooperation. Reproductive cooperation is an attractive option as the foundational form of hominin cooperation because it was stable. The fitness interests of the key maternal supports—grandmothers, sisters, children's siblings, fathers— are in important ways aligned to the fitness interest of the mother. Hunting coalitions and social learning networks are not typically bands of brothers, so we need to explain the resilience of ecological and informational coop- eration, given the apparent threat of free riding. To that I now turn.

# 5 Costs and Commitments

## 5.1 Free Riders

I have argued in the last four chapters that one of the most distinctive features of human social worlds is our dependence on intricate networks of cooperation and the division of labor. No living humans gather the resources needed for a successful life by their own efforts. None do so even with the help only of their immediate family. Yet it is a truism of evolutionary biology that cooperation evolves only in special conditions. While cooperation can be immensely profitable, it is also fragile. In many circumstances, the profits of cooperation do not depend on all who benefit paying their share of cooperation's costs.[1] Hence many potentially profitable cooperative partnerships cannot establish, for they would be destabilized by cheating. It may be that cooperative collapses have caused local extinction, and perhaps more often the extinction of identifiable cultural groups. Yet our ancestors cooperated, and we continue to cooperate. Cheats exist. But the costs imposed by cheating have been kept low enough to make cooperating a successful strategy.

In section 1.2, I discussed the hypothesis that the dilemmas of cooperation drove the evolution of hominin intelligence via an unstable interaction between the temptation to seek the benefits of cooperation and the threat of cheating. We are intelligent because cooperation is at once risky and too profitable to abandon. Cooperation demands careful management, for cooperating with defectors is costly. Likewise, unsuccessful attempts to defect are often expensive. We are contingent, judicious, and wary cooperators (Humphrey 1976; Flinn et al. 2005). This perspective on hominin cooperation somewhat mischaracterizes the evolutionary dynamic (Sterelny

2007). According to this Machiavellian view, cooperation management is cognitively demanding because it is difficult to identify cheats.[2] Cheats have an interest in seeming to be reliable cooperators, and their mimicry is difficult to unmask. To detect them, we depend on our most sophisticated capacities, for example, on forensic uses of our theory of mind.

It is true that detecting cheats in mass society is indeed cognitively difficult, for we will often be interacting with strangers. But in small-scale, intimate social worlds, our ancestors were awash with information about one another. Partners in collaborative foraging activities are bound to find out a lot about one another as a side effect of their mutual economic activities and through intimate association in stressful, tiring, and sometimes dangerous activities. It is hard to keep a secret in a village. In small-scale social worlds, the problem is not detecting cheats; the problem is controlling them, for effective sanctions are rarely free.

One important mechanism that reduces the impact of cheating is partner choice. Many of the formal models of cooperative interaction focus on the problem of control: how can one agent influence the payoffs of his or her partner in ways that induce or encourage that partner to respond cooperatively? In this spirit, many strategies of contingent cooperation have been explored: strategies in which one player will cooperate contingent on the past choices of other players. But as Ronald Noe (2006) argues, in many circumstances, successful cooperation depends on identifying reliable partners while being yourself reliable, for many collaborations begin with mutual choice. In such cases, the problem is not to force cooperation from those with whom you are fated to interact. Rather, it is to choose well in the first place, and to be chosen by others. For example, agents (or their families) often have considerable influence over whom they marry. But once they are paired, local practice will often determine the shape of the interaction: the residence pattern, division of labor, initial resource contributions, costs of decoupling. Obviously in marriage-like arrangements each partner has some ability to shape the behavior of the other. But much hangs on partnership choice. In most small-scale, traditional social worlds, that choice is informed by local reputation. Even in those informationally rich worlds, wise choice may not be easy. For one thing, many cooperative alliances involve long-term commitments: selecting partners who will continue to be reliable over changes in circumstance and individual life history. While partner choice plays a central role in explaining the evolutionary stability of human cooperation, future-proofing cooperative ventures by wise choice of partners is not easy.

Moreover, cooperation cannot always be managed by wise partner choice. Agents cannot always choose their partners in social interactions. Without choice, cooperation does depend on partner control. The problem is particularly vivid when defection consists not just of passive free riding but active and aggressive exploitation, where bullies seek to appropriate collectively produced resources. Bullying is radically destabilizing. Suppose one agent spends time and effort finding food, making a tool, building a shelter, starting a fire. What prevents such an agent from having his reward routinely appropriated by any physically stronger group member? Nothing remotely resembling a human social life could evolve until such investments were worthwhile. Yet there seems to be a temptation to seize such resources. Indeed, the more another agent has added value to collected materials, the greater the temptation. Elementary, and presumably ancient, forms of human cooperation depend on the fact that resources were not routinely seized by the strongest individual in the vicinity. But while exploitative bullies are all too easy to identify, they are not easy to control.

## 5.2  Control and Commitment

Destabilizing bullies are likely to be physically dangerous, so the costs of control are high for single individuals unfortunate enough to be in a bully's sphere of action. The only individuals powerful enough to punish a bully without excessive risk would maximize fitness by becoming bullies themselves. So control must be exercised collectively, and when a group acts as a committed, coherent coalition to punish greed or free riding, it can do so effectively and at low cost to each member of the enforcing coalition (Boehm 1999; Bingham 2000; Boyd et al. 2005). Male chimpanzees form coalitions and in the security of a coalition are able to inflict deadly violence on lone males without risk (Wrangham 1999). I think it likely that early hominins also had coalition-forming capabilities, for as they found themselves exposed in increasingly open habitats, they would have been subject to strong selection for cooperation against predators (see sec. 4.1). The evolution of defensive or offensive coalitions will put in place some of the cognitive, motivational, and communicative mechanisms needed for the collective suppression of bullies. But not all. Such coalitions form and act against outsiders; they do not form in the face of within-group suspicion and disruption. The formation and smooth operation of a coalition cannot be guaranteed, especially in the face of attempted disruption and intimidation by aggressive and dangerous bullies. If everyone in the potential coalition is resolute and acts together, no individual will be able to stand against it. But if coalition formation misfires, the attempt at punishment can expose an agent to deadly retaliation (Sterelny 2003). The potential costs of punishment, then, are high, so there is a problem in making the threat of punishment credible. So it is likely that an early and important form of cooperation consisted of cooperative threats—threats to punish as part of a coalition—and we need to explain the credibility of those threats to both allies and their targets.

The control of antisocial behavior by threat of punishment exemplifies the so-called commitment problem (Frank 1988, 2001; Schelling 2001). A commitment problem is structured as follows:

An agent optimizes his (her) future utility if (s)he acts at time $t$ to ensure that at $t + n$ (s)he will carry through option $X$ (if some triggering event occurs), even though choosing $X$ at $t + n$ does not then optimize utility. The act or activity at $t$ that ensures $X$ will be chosen at $t + n$ is a *commitment device*.

Committing yourself at $t$ to make a suboptimal choice at $t + n$ seems paradoxical but is not. In the case of threats, your commitment lowers the probability of the triggering event occurring. The commitment to punishment makes it less likely that the agent will have to defend his resources against seizure, and less likely that he will need to punish successful theft. By making the threat of retaliation credible, even though it would then be expensive to strike, an agent secures serene possession of his resources. It is rational to defect in a classic, symmetrical, one-shot prisoner's dilemma. But it is also rational for two agents to commit to each other to cooperate should they ever find themselves about to be paired in a classic, symmetrical, one-shot prisoner's dilemma.

Problems with a similar structure arise in many cooperative enterprises. For example, Smith and Bliege Bird note that a trust problem often arises in long-term cooperative partnerships:

In choosing a partner . . . such as a mate to cooperatively raise offspring, a research colleague, or a co-author—each partner must be convinced that he or she will gain a net benefit. . . . This is particularly important if the association will be a long-term one, with opportunities for cheating or periods of one-sided costs . . . that would allow the other partner to maximise short-term gains by defecting. (Smith and Bliege Bird 2005, 139)

Thus commitment devices are important to partner choice, not just partner control by credible threat. Partner choice really does help explain the stability and prevalence of hominin cooperation. But cooperation through wise choice depends on the existence of commitment mechanisms. If an agent can credibly commit to continued cooperation even if the opportunity for profitable defection arises, agents secure cooperative partners and the profits of cooperation. If they could not commit to cooperation even when faced by temptation, they would not build the cooperative partnerships that generate tempting options. Agents face temptation only if they credibly guarantee that they will not succumb.

In short: if agents can commit—if commitment devices are available—agents can enhance fitness by constraining future choice. Constraining future choice enhances an agent's capacity to deter. Credible threats secure resources that enhance fitness. Constraining future choice also enhances trustworthiness. Trustworthy agents can enter extended, profitable partnerships that cannot be stabilized by mutual surveillance. Since the profit of cooperation often depends on a division of labor, mutual surveillance is often impossible. Without surveillance, cooperation requires trust. But how is commitment possible; how do commitment devices work? Agents need to be able to bind themselves to future commitments and to signal that they have thus bound themselves. To solve, say, the deterrence problem, an agent must be able to bind himself now to retaliate in the triggering situation, and he must be able to signal that fact to an audience. As Dr. Strangelove so emphatically pointed out, there is no point in building a *secret* doomsday bomb. Yet how can an agent rigidify his current intention to retaliate? If deterrence fails, retaliation is more costly than it is worth, even by the agent's own lights. How can those who receive a signal discriminate between the truly committed and the bluffers, given that it is in an agent's interest to seem committed, even if he is not? In this picture, warranted trust in threats and promises explains how cooperative alliances are formed and how cheating is controlled. It explains why cheating does not corrode collective action. Warranted trust depends on commitment. So we need a theory of how commitment devices work and of how they can be assembled by evolutionary processes.

## 5.3  Commitment Mechanisms

In Schelling's classic identification of the problem and profit of commitment, he distinguishes between external and internal solutions to commitment dilemmas (Schelling 2006). In external solutions, the committing
agent acts on his environment to change the payoff matrix in the trigger
situation at $t + n$ to remove the temptation to renege on his commitment.
The agent's intervention ensures that maintaining commitment is the optimal choice at the $t + n$ trigger. In the standard illustration of this idea, a
general burns the bridges behind his invading army, making retreat truly
catastrophic. Even if the battle goes badly, fighting on is the least bad option for all, and thus the general increases the prospects of victory by increasing the costs of defeat. External solutions to commitment dilemmas
are also available in more mundane cases. In signing contracts, in under-
taking the formal and public rituals of marriage, in publicly staking one's
reputation on a course of action, an agent changes the payoffs for failing
to cooperate at a triggering temptation. The price of acting on temptation
increases, perhaps dramatically.

However, received wisdom suggests that these external solutions typically depend on the institutional and cultural framework of contemporary
and near-contemporary life. Now and in the recent past, external solutions
to commitment problems are often available. An agent can bind himself to
matching his commitment by changing the payoffs of failing to do so. Moreover, this commitment can be signaled unequivocally. We and our recent
ancestors live in a world with standard, institutionalized legal contracts,
rank and command systems in armies, marriage ceremonies and similar
public rituals of commitment, systems of accreditation in professions, and
the like. These social institutions make it possible for agents to change their
payoffs at triggering situations, and make it possible for all to recognize that
a commitment has been made. However, such institutions emerge only in
an environment in which elaborate, ongoing cooperation has already been
established. Commitment dilemmas were regularly solved before, and as
a precondition of, the cultural evolution of institutionalized frameworks
that enforce commitment. Enforcement coalitions and other cooperative
interactions must have been an ancient and foundational aspect of human
social life; they formed a cooperative foundation on which the institutions
of complex society were built. In explaining the cooperative bedrock, we

cannot appeal to institutions such as formal laws or institutional structures of command and control.

These supposed limits of external solutions explain Frank's classic exploration of internal solutions to commitment dilemmas. Internal solutions depend on the agent's own psychology (Frank 1988). Internal commitment devices reevaluate the options at $t + n$ by adding internal costs and benefits to the payoffs of different actions. Reevaluating utility makes breaking faith less tempting. Since the psychology that mediates reevaluation is as old as our species (perhaps older), commitment dilemmas could be solved by humans who built the cooperative fabric of contemporary human worlds. Internal mechanisms, in this view, were foundational to the solution of commitment dilemmas and hence to the evolution of human ultrasociality. In particular, Frank argues that our distinctive social and moral emotions are commitment devices. They cause us to evaluate our options at the commitment triggers, so that our subjective assessment of utility matches the option to which we have committed. As a committed agent evaluates options, the utility-maximizing act is the trust-keeping act, whether or not it maximizes resource take. As it happens, in hominin social worlds, the economic rewards of being trustworthy are important. Agents who care about keeping commitments tend to optimize their long-run economic welfare. As increasing resource take tends to increase fitness, there is selection for being trustworthy. Agents are trusted only if they can credibly commit; agents who can credibly commit gain an advantage thereby, and that explains the evolution of the commitment emotion complex. Commitment dilemmas were important to hominin social life. And so we evolved emotions that are motivationally powerful, emotions that are triggered by perceived violations of trust and fairness, emotions whose motivational saliences are relatively insensitive to utilitarian calculation, emotions whose occurrence are easily recognized and difficult to fake (Nesse 2001).

For example, imagine an agent who knows that it will cost him a fortune in time, energy, and money to nail the guy who fender-bent his car. He knows it does not pay economically to punish. But he really does not want the cheat to get away with it. As he assesses the likely outcomes, the subjective satisfaction of successful punishment boosts the expected utility of punishment, and the subjective dissatisfaction of the cheat getting away with it increases the cost of letting the incident go. As the agent weighs the costs and benefits through the lens of his social and moral emotions, seeking punishment after deterrence has failed is not irrational. It is utility maximizing, though the utilities thus maximized are not economic returns. These emotions change patterns of social interaction by causing some agents to heavily weigh loyalty in cooperative interactions, fairness and honesty in bargaining, and getting even in response to defection. Others, recognizing these emotions, trust them.

Frank's idea faces two challenges. First, he must show how emotions function to motivate committed choices, though they do not maximize resource gains, and though agents are not and cannot be motivationally indifferent to resource rewards. Indeed, humans (and other animals) have great difficulty resisting immediate reward, even when deferring gratification would pay. That said, experimental economics suggests a picture similar to the one that Frank develops. In these experiments, agents interact, typically anonymously, in cooperation and defection dilemmas. For example, in public-goods games,[3] agents can keep their private stake or contribute from it to a collective pot, where its value grows before the pot is distributed evenly among all. The selfish strategy is to keep your private stake while hoping that others contribute to the public fund. But often that is not what we see, and these experiments seem overall to show that agents are "strong reciprocators." They are cooperative; they expect others to cooperate, but they are vengeful if they encounter defection, sometimes even when they are not the injured party (Bowles and Gintis 2003, 2004; Fehr and Fischbacher 2003, 2004; Gintis et al. 2005). It is important to note that experiments in this genre have been conducted cross-culturally, including subjects from small, traditional societies. While we see important variation from culture to culture, the pattern remains broadly similar (Henrich et al. 2005).

The idea that humans are strong reciprocators is controversial, and I return to this issue in sections 8.2 and 8.3. But let us suppose that some agents are indeed trustworthy by virtue of their emotional structure. A second problem arises. In a world in which others are willing to trust, an agent who is not trustworthy would gain from seeming to be trustworthy. So how can a committed agent *signal* trustworthiness in a deceptive and thus skeptical world? While I think that Frank's model provides important insights, I argue that commitment mechanisms primarily depend not on subjective rewards and signals but on constructing environments that channel choice. The signaling problem is not quite so pressing as Frank's model implies. Agents induce themselves to keep faith by reducing temptations to break faith. One important commitment mechanism is niche alteration. Reevaluating outcomes—placing value on loyalty, fairness, or revenge—is important too. But these subjective factors do not and did not carry all the weight. Hybrid solutions to commitment problems, solutions involving external, objective elements, are ancient.

## 5.4   Signals, Investments, and Interventions

If trustworthiness depends on the emotional makeup of agents, trustworthiness cannot be observed directly. An agent must somehow show, display, or signal her trustworthiness. Since it is in the interests of the less trustworthy to seem more deserving than they really are, there is a deception problem. One response has been to exploit the costly signaling framework developed in section 4.5. Costly signals are honest signals of sender characteristics, because only honest signalers can afford the handicap of sending them (Grafen 1990; Zahavi and Zahavi 1997; Searcy and Nowicki 2005). The basic structure of a commitment dilemma seems to fit the costly signal framework. Trustworthiness is not a directly observable trait; it needs to be *displayed*. Other agents have an interest in recognizing the display; less trustworthy agents have an interest in faking it. In addition, the idea that humans enforce honesty in social negotiations through signal cost has real ethnographic traction, for the anthropological record is full of costly or dangerous practices that seem to serve no economic interest but are naturally interpreted as signals (Bliege Bird and Smith 2005). Think of bodily adornment and alteration; rituals and ceremonies; adolescent chicken games and similar gratuitously dangerous acts; decorating utilitarian objects; producing objects, structures, and buildings with no utilitarian function; and conspicuous consumption—the consumption of resources that are attractive only because they are rare and expensive.[4] As we saw in section 4.5, even the apparently economic activity of hunting has been interpreted as male display.

  I agree that the costly signal framework explains the dynamics of some human interactions, but its scope is narrower than its proponents suppose. The costly signaling framework does not apply well to systems in which most members of the relevant population signal in similar ways yet signal honestly. The role of cost is to make signals, or high-intensity signals, unaffordable to many in the population. Costs expose outliers. Yet many high-cost human activities seem to be one-in, all-in systems. They involve all members of the group signaling. Initiation rituals, for example, often impose high costs on the whole age cohort of a group. And the cost falls equally on each: members of a cohort do not compete by varying signal intensity; often they are pass–fail systems (for an entertainingly presented set of examples, see Barley 1986). Universally imposed costs cannot discriminate among those paying the cost, but they can be part of a commitment mechanism. Not every member of the group can be of higher-than-average quality. But everyone's threats or promises can be credible. Trust and trustworthiness need not be a scarce resource (though obviously if trust is not scarce, its relative value declines). However, though I do not think costly signal frameworks apply well to commitment dilemmas, commitment mechanisms do depend on costs, in three ways.

First, *costs amplify effects of arousal*. Some commitment signals are motivation bending rather than information carrying; they are Krebs-Dawkins signals (Krebs and Dawkins 1984). Their function is to change the motivational psychology of both sender and receiver rather than to reveal antecedently existing characteristics of the sender. Singing, for example, is a signal. But it has an affective impact on both sender and audience (Mithen 2005). Signal costs, I suggest, up-regulate the effects of such mood- and emotion-altering signals. Costs help solve commitment problems by increasing the potency of such signaling. I presume this connection between cost and arousal is ancient, long predating the evolution of human cooperation. By amplifying the salience of both reward and punishment, reinforcement is more effective in highly aroused situations. So joint action in emotionally amplified situations reinforces mutual bonds more powerfully than collective action in calmer emotional waters. I argue in the next section that to the extent that hunting is a signal at all, it is a Krebs-Dawkins signal, and shared dangers and shared success increase the power of mutual manipulation.

Second, *costs are investments*. We commit through niche alteration. But changing the world is not free. In these cases, costs are the direct or indirect price of changing the environment, so that carrying through the commitment becomes the optimal rather than the suboptimal choice. These investments also function as signals, because the niche alteration is public and is intended to be public; the investments change the agent's payoffs in ways that are apparent to other agents as well. Tattoos and facial scars are not (just) signals but interventions. They make cooperation within the group the right option in almost all circumstances, for the tattoos make shifting social networks much more difficult. In advertising your origins and affiliations, you inherit your allies' enemies, whether or not you keep your allies' support. You are trustworthy because you have no other choice. So these signals constrain choice, and constrain it publically, rather than reveal agent quality. I discuss commitment through investment in section 5.6.

Third, *honesty has a by-product advantage*. Honest signals can take advantage of by-products that increase their credibility for free. A dishonest signaler has to manufacture the evidence that makes a signal credible as well as produce the signal itself. Honest agents often rely on cues rather than purpose-built signals for reputation benefits. Deception can never rely on cues; a deceptive agent must pay to send a signal. This generates an automatic cost asymmetry between honest and deceptive signals, one unconnected to signaler quality. It is easy and cheap for me to present credibly as an Australian birder, because I am an Australian birder. The resources needed to make my birding credentials credible have been assembled automatically, as a side effect of my birding, and independently of any need to signal that characteristic. Becoming an Australian birder was not effortless. But once I became a moderately competent birder, signaling that trait became effortless. Indeed, often I do not need to signal; this information leaks from my ordinary life activities. Suppose someone with no birding skills wanted to credibly impersonate being an Australian birder. That would impose high signal costs. They would have to outlay funds on peripherals: optics, field guides. They would have to spend time and energy acquiring at least a passing knowledge of the birds, especially if they could not rely on an utterly ignorant audience. It is possible to fake birding interest and expertise, but it would be expensive. It is cheap for me and expensive for a fraud. If the profit of deception is low, this asymmetry in itself might suffice to keep signals honest.

This cost asymmetry is a structural feature of signaling situations. But it can be amplified if skeptical receivers scrutinize signals for independent confirmation, or if senders repeatedly interact with receivers in slightly different contexts. Equally, it can be reduced if receivers are credulous, or if the deceptive agent only has to pass muster once. But the mechanism itself does not rely on any underlying difference in the quality between the honest and the deceptive; the one agent can be honest in some contexts and deceptive in others. The difference in relative cost is driven by the general fact that faking a property always depends on signal production. It is never an informative by-product of other behavior, a by-product that can be co-opted as a cheap signal or as evidence that makes a signal credible. Thus faking involves a cost that honesty does not (Bacharach and Gambetta 2001; Gambetta 2005).

Robert Frank himself relies on this third mechanism of signal honesty, though he also suggests that cues have become elaborated as they have become co-opted for communicative functions. Commitment emotions involve complex displays. A convincingly honest smile while making a credible bargain involves body posture, complex facial motions, shared gaze, voice. Someone who is honestly promising generates the complete syndrome as an automatic consequence of the emotion she is experiencing as she commits to keeping the bargain. A smile is probably a signal rather than a cue, but an honest smile is a low-cost signal. A fake would have to invest considerable cognitive and developmental resources to bring this syndrome under top-down, voluntary control. Actors can learn to produce "honest" smiles, but it is not effortless. Other commitment emotions may be even more difficult to simulate. Dan Fessler points out that not even the best actors produce real laughter on demand. So he thinks it is no accident that many cultures have norms about bargaining that involve social activity, joke telling, and laughter. These norms license probing the emotional attitudes of those involved in the deal in ways that are difficult to fake for those who have entered into contracts in bad faith (Fessler and Quintelier, forthcoming). Thus an initial cost asymmetry is magnified by social engineering. Stalin's late-night drinking sessions with his inner circle exemplify such probing in its most extreme form. Stalin's elite were repeatedly scrutinized in the early morning hours, and for hours, by Stalin himself, completely sober, while they were relentlessly plied with alcohol (refusal was itself dangerous). They were, of course, acutely aware of the fatal consequences of any suspicion (Montefiore 2005). So they were in circumstances of exhaustion, stress, and deinhibition, conditions under which it would be ferociously difficult to retain top-down control of emotional signals.

Stalin built on an initial cost asymmetry by leveraging agents into environments in which top-down control was appallingly difficult. Frank thinks that evolution has made top-down control difficult by elaborating the expressive syndrome of emotions. As he sees it, the complexity of human emotional expression and the increased costs that complexity imposes on faking is an evolved response to the honesty problem in commitment signaling. Emotional displays might have become more elaborate to increase the investment needed to fake commitment emotions (thus Fessler thinks that laughter has only recently been recruited as part of a commitment signaling system, and that is why it cannot yet be convincingly simulated). A commitment itself—say, to deter—may end up being expensive if the threat itself does not suffice. But when it is honest, the signal itself is cheap, much cheaper than a fake signal, for it requires none of the scarce cognitive resources of top-down attention, control, and self-monitoring. In Frank's picture of the role of the emotions, cost is relevant to success in signaling commitment by increasing the cost of fake signals. But the mechanism does not act via a handicap principle. An honest signal is cheap and carries information about the reevaluation of options, not agent quality.

## 5.5 Hunting and Commitment

Perhaps the only consensus on the role of large-game hunting in forager life is that no single model fits every case. Some costly acts are signals, and some costly signals fit the handicap model. But despite the Hawkes-O'Connor analysis, forager hunting is rarely well explained as a costly signal of male quality. Return rates are high, and sharing is contingent on reciprocation and relatedness. So much hunting *is* well explained in economic terms (see sec. 4.5). Moreover, the phenomenology of male hunting fails to fit costly signal predictions, for crucially, in many forager worlds, all males hunt: it is a central preoccupation of their lives. Hawkes and her colleagues insist on this fact themselves, and for good reason. The cultural importance of hunting to men makes it plausible to claim that hunting is not purely instrumental. But it does not fit the costly signal framework either, for the standard expectation of costly signal models is that only high-quality males signal (or that they signal in proportion to their quality). If the hunting return rate is the signal, low-quality males pay the costs of signaling while advertising only their incompetence. They should either abandon hunting for individual resource acquisition, or hunt without sharing, or hunt much less (perhaps only when they are especially favored by ideal conditions). Universal signaling is at odds with the handicap framework.

As we saw in sec 4.5, the idea that only high-quality individuals send costly signals is a little too simple. In some versions of the framework, agents signal with an intensity that varies in proportion to their quality. These are called "revealing" handicaps (Searcy and Nowicki 2005), and Robert Frank (1988) has argued that such signals are governed by a "full disclosure principle." This principle can explain honest signals in contexts of conflicting evolutionary interest, even when a signal honestly advertises the frailties of a less-than-ideal agent. The full disclosure principle explains, for instance, why agents produce shifty smiles rather than inscrutable poker faces. Frank explains the full disclosure principle via an idealized model of frog croaking. Females prefer large males, and the physiology of sound links croak pitch to frog size. Deep croaks mean large frogs: it is not physiologically possible for small frogs to produce deep (as opposed to loud) croaks. Given female preference, it is no puzzle that the largest frog croaks. But what about the rest? Why does a smaller frog keep croaking once he hears he is not the largest? The second-largest frog (Frank explains) should croak too, for he can advertise the fact that he is larger than the average of the rest; and should he keep silent, that will be the female's default estimation of his size. By parity of reasoning, the third largest should croak too, and so on for the rest. The basic mechanism, Frank thinks, is that failure to display will generate an audience estimate that understates your quality, at all stages of the dynamic. A signal-or-we-assume-the-worst dynamic plus the physical constraint on signal production generates an informationally transparent environment. Despite their conflicts of interest, all relevant agents signal, and signal honestly. Honesty here depends not on cost but on the existence of a trait that is an index of quality by physiological necessity. High-quality agents are selected to display that trait, to make it perceptually salient, and this drags the rest along. Since there is genuinely important information to be had, the audience is selected to attend to the signaling.

Alas, unsurprisingly, the real world is not so benign. For one thing, there is rarely a physical constraint that welds a perceptually salient property unbreakably to a biologically salient property. We have certainly had no such luck in the connection between genuine smiles and promise keeping. More generally, the full disclosure principle neglects costs. Croaks are energetically expensive, and they attract predators as well as female frogs. So a tipping point will come where the costs of croaking outweigh the costs of suspicion, and disclosure becomes too expensive. As a result, real frogs (and animals with similar display signals) have evolved a number of non-signaling sexual strategies: female mimicry, ambush mating (small males that cluster near large croaking males to attempt opportunistic fertilization), and the like.

The bottom line, then, is that neither the costly signal framework nor related frameworks are well poised to explain costly but universal signals. So I have an alternative suggestion: hunting is a commitment signal, and it builds commitment more effectively as a result of its costs. Hunting signals to others in the band the agent's loyalty and reliability. Hunting together builds trust. Hawkes, O'Connor, and their colleagues build a persuasive case for the idea that hunting has a psychological and cultural salience in many male foragers' lives that depends on more than its return rate. Perhaps this salience results from hunting's role in building commitment in male networks. A social network of partners, each of whom trusts the other, is a crucial life resource. One role of hunting in many forager cultures is to build that trust through shared activity and experience. Hunting together generates internal and external factors that increase the probability of each trusting the others, and of each being trustworthy, thus initiating positive reinforcement loops.

In my view, both internal and external factors are important to the role of hunting in commitment. Beginning with internal factors, Michael Tomasello and his lab have drawn attention to a strikingly, and possibly uniquely human, phenomenon of shared, joint intentions (see, e.g., Tomasello et al. 2005; Tomasello and Carpenter 2007; Tomasello 2008, 2009). A shared intention is partly cognitive: agents in a collective activity have mutual knowledge of one another's attentional focus and activity. Indeed, the human eye is adapted to facilitate joint attention, by making our direction of gaze evident. Our scleras are white and enlarged and contrast with our dark pupils. In contrast, great apes' eyes have colored scleras, probably to reduce contrast and disguise gaze direction.[5] But joint attention is partly a motivational state as well. In their experimental work, Tomasello and his colleagues show that children (but not young chimps) engage in these joint activities *because* they are joint (Warneken and Tomasello 2009; Warneken, forthcoming). Humans are cooperative in part because we find collective activity rewarding.

Collective hunting is not just a joint activity: it is often an extremely demanding joint activity. The psychology of team sports, and the regular reports in memoirs of line soldiers, show that shared danger and stress in collective activity are immensely affectively powerful (see, e.g., Graves 1929; Keegan 1983; L. MacDonald 1993, 1994; Fraser 2007). Soldiers' memoirs are particularly telling: they universally emphasize the interaction between group loyalty and combat. The intense experience of combat both generates deep loyalty and can be sustained only because of those ties of loyalty. If hunting is like combat, the mutual experience of hunting and sharing does not signal *a preexisting bond* of loyalty and attachment that makes trust and trustworthiness likely. Rather, it incrementally creates those psychological bonds. Trust and trustworthiness are both the products of costly, high-arousal activity.

Importantly, this mechanism need not discriminate within a group of interacting agents, marking some as being of higher quality. Trust is not like status: it need not be a limited resource that is reserved only for the stars of the hunt. All members of a group can show themselves worthy of trust: they can show that in situations of stress, danger, or fatigue, they will do what they can for others. A good team player need not be the star of the team. The paradigm examples of costly signals are prey-to-predator signals and male-to-female signals. These are zero-sum interactions. If one agent increases his probability of sexual access by advertising his high quality, he does so at the expense of the prospects of some other agent(s). Commitment building need not have this zero-sum character: if I am now more trusted, it does not follow that trust in another has been withdrawn. So the role of cost need not be to exclude all but the best, making the cost of advertising unaffordable for all but a few.

Cost in part has an epistemic function: it is harder to fake a role when you are tired, hungry, cold, and scared; as we saw earlier, Stalin was acutely aware of these unmasking effects. But costs also increase arousal, making the effects of joint activity more powerful at a time and over time. These costly collective activities change the agents who share in them in relevant ways; they do not just advertise preexisting capabilities. Defenders of the costly signal framework downplay the importance of this effect. For example, Hagen and Bryant (2003) have argued that song and ritual are costly signals of alliance quality: an elaborate, coordinated, cohesive display is an impossible-to-fake signal that those displaying have spent time and energy together. They are an established collective with a real joint history who can act together. Quite explicitly, Hagen and Bryant treat ritual and song as a purely informational signal: these activities carry information about preexisting characteristics of the individual and group. While they briefly mention that ritual and song have emotional and cognitive effects on those engaged in such joint activities (and on the audience), the authors do not think of these effects as part of the function of ritual and song. That seems to be an implausible view of ritual and song—these signals may be informational, but they are also Krebs-Dawkins signals. An analogous view of collective hunting is almost as implausible. Participating in collective hunts reworks the emotions both of the agent and of his partners.

In this view, the costs of joining collective actions are indeed relevant to commitment: they tune up the affective amplitude of the interactions. But they also play an economic role in securing commitment; external factors are also important. The more the male network builds mutual trust through the collective experience of hunting and sharing, the greater the benefits of that network to each of its members. Consider foraging lives like those of the Ache. Individuals take part in both individual and collective hunting, and members of foraging parties, in particular, share quite generously (Gurven 2004; Allen-Arave, Gurven, and Hill 2008). In such worlds, trust has been built incrementally. Individuals have invested time, energy, and resources in specific relationships. In all probability, no specific hunting trip has imposed a major cost (unless one agent has rescued another at real risk). But the total investment that individuals make in building relationships is large.[6] When all goes well, the investment secures an important asset: partners who can be relied on for social and economic support. But that asset is not transportable, nor can it be readily transformed into a transportable asset. Investment in building and maintaining relationships is a commitment device, because such relationships change the payoffs in triggering situations. If a triggering temptation arises, the costs of defection have been driven up. Defection will risk fracturing trust and hence forfeiting the profit from the investment necessary to build trust. So defection may well no longer be the objectively optimal decision. Finally, third-party effects are important. As the capacity to communicate about events displaced in time and space comes onstream, reputation becomes important as well. Both keeping and breaking faith will have third-party as well as second-party effects. In short, commitment by investment is an important mechanism. It is the focus of the next section.

## 5.6  Commitment through Investment

These ideas about the dynamics of hunting parties show that one solution to the commitment problem is to act on the environment so as to alter its cost-benefit structure, eliminating the temptation to defect (Frank 1988, 2001; Schelling 2001). Investment makes defection less tempting; it changes the payoffs in the option pool. Smith and Bliege Bird (2005) illustrate this idea through the charming example of showing commitment by ceding first authorship on an important paper or by inviting coauthorship in a book project that one might have kept to oneself. The signal is costly because it is an investment. As such, it changes the payoff structure. Once one researcher has invested coauthorship in the partnership, she has less temptation to defect; she would lose her investment. Such examples show that some external solutions to commitment problems do not depend on Holocene institutions.

A committing agent can choose a pattern of investment that will be individually profitable if and only if the commitment is kept. The agent can expect a profit only if he resists the temptation to defect should a triggering situation occur. These investments are typically public and readily observable. So the alteration in the payoffs an agent faces in triggering contexts is known to others. The commitment mechanism is not a hidden internal characteristic, requiring extra machinery to reveal. Accumulated investments in the relationship raise the costs of failing to carry through on a commitment, making trust by others more rational. The availability of external but noninstitutional solutions to the commitment problem is important. If, for example, the members of a punishment coalition regularly cooperate in minor ways, in interactions that generate individually small but numerous profits, each has a lot to lose by forfeiting trust. If agents cooperate together in different ways, each makes the others more stable by increasing the benefits of inclusion and the costs of exclusion, and by reinforcing through repetition the affiliation built by shared successful activity. External factors reinforce internal ones. Moreover, since such agents have a history of successful coordination, the informational challenge of coordinated action in risky circumstances is eased too. Such practiced agents will be sensitive to subtle cues of one another's intentions, just as long familiarity in playing together eases coordination in team sports. We do not have to suppose that foundational solutions to commitment dilemmas depended on intrinsic psychological mechanisms that devalue the temptations of material rewards and increase the value of affective rewards. We do indeed seem to have such mechanisms, as the literature of strong reciprocation shows (sec. 8.2). But they may have evolved in response to new social worlds of extended cooperation, and in conjunction with niche-altering commitment mechanisms.

A coevolutionary perspective is crucial to understanding information sharing and cultural transmission, and it is also crucial to understanding commitment. The commitment mechanisms just identified did not evolve, and do not operate, independently of one another. Commitment acts often are partially Krebs-Dawkins signals, partly investments, partially option reevaluations. These cost-based mechanisms can reinforce one another. The hunting and collective-action cases I have been discussing are hybrid solutions to commitment dilemmas. Moreover, the external, investment-mediated factors operate incrementally as a group builds trust through a pattern of joint high-stakes action and mutual investment. The network of mutual trust emerges gradually. Successful cooperation is self-reinforcing, and these reinforcement mechanisms do not depend on contemporary levels of individual cognitive sophistication or on the elaborate social networks and institutional structures of contemporary and near-contemporary life.

When a group exists prior to an individual who wants to be part of it, the relationship is asymmetrical; the commitment mechanism is more purely external and demands from the committing agent a more dramatic up-front investment that reshapes his or her payoff in triggering conditions. In contemporary social environments, gang initiation rituals often have this character. They often involve actions that indelibly physically mark an agent, rendering his presumptive affiliations obvious to friend and foe alike. And such initiation rituals often involve high-price defection from the norms of the larger society. Thus apparently one motorbike gang in New Zealand requires an act of rape to secure full membership. If so, a patched member of the gang cuts himself off from success by the ordinary avenues of New Zealand society. He can succeed only in the gang, and with its cooperation. Remaining staunch in the face of temptation no longer requires ignoring the optimizing choice. Assuming a normal weighting of outcomes, it *is* the optimal choice. The macho swagger and intimidating gestures of the archetypal gang member are probably Zahavian signals, for such acts impose risks that only the genuinely dangerous could survive. But the tattoos and patches are not Zahavian costly signals. Though the mechanism depends on costs, it does not depend on the idea that those costs fall differentially on individuals of different intrinsic quality. They need not be handicaps that only the truly feral can afford. The insignia of membership and exclusion are public investments that change payoffs in the triggering context in ways that everyone will recognize. Fessler and Quintelier (forthcoming) detail American prison examples that make the same point about the role of gang initiations and insignia. They make it suicidal to defect from a gang, once joined, because they mark you publically and indelibly as a target to other gangs.

Gang initiation rituals depend on the institutional framework of a contemporary liberal Western society. But the general mechanism does not. Many of the initiation rituals discussed in ethnography are likely to be commitment devices that work by making defection costly.[7] For example, many initiation rituals involve the imposition of scars or tattoos that unmistakably and indelibly advertise group membership to friend and foe. Once a Maori boy has the full moko, there is no denying his origins. By making migration between groups more difficult, such signals make individuals within groups more trustworthy; they must hang together or die separately. Indeed, Dan Fessler suggests that ethnographic evidence shows that aggressive, warlike cultures often impose highly visible, permanent marks on their members, marking them as belonging to a feared and hated group, binding their fate to that of their group.

The moko and similar examples work by decreasing mobility and thus increasing the agent's stake in his or her current network. Other noninstitutional investments have a different dynamic. Consider, for example, the widespread and clearly costly practice of female genital mutilation.[8] The origin and establishment of female genital cutting is not known, but it is naturally interpreted as a commitment mechanism, one that invades because it gives fitness advantages to early adopters. It is then stabilized by the fact that once the practice is common, unilaterally abandoning it becomes individually expensive. Female genital cutting thus establishes as an external commitment mechanism. A poor family can marry up if the bride can send an unmistakably honest guarantee of fidelity. Female genital cutting is such a mechanism because it removes (or so it is believed) the wife's temptation to sexual defection. By undergoing female genital cutting, she alters her future payoffs in commitment situations. She can buy her way up, thus improving, at a price, her own prospects or those of her family. So female genital cutting may well have begun as a commitment device, giving an initial advantage to early adopters in a marriage market. But once this practice is copied, it becomes a default. Girls can no longer improve their position relative to others by female genital cutting. They can only penalize themselves (perhaps massively) by abstaining from it. So once established, individuals cannot profitably defect from this convention, even when most of them would like to do so. A striking fact about female genital cutting is that many who practice it, and intend to continue practicing it, disapprove of it. Undergoing female genital cutting has become so important to marriageability in the local environment that individuals, even when they disapprove of the practice, cannot defect from it without severe penalty (Mackie 1996; Bureau 2001).

The same pattern seems to fit foot binding. Foot binding, like female genital cutting, imposed clear and obvious costs. Since it involved crushing the feet of small girls, it was appallingly painful for them and severely restricted their mobility when they reached adulthood. The restriction of mobility makes foot binding a chastity commitment device: the temptation to defect is much less likely to be available. Foot binding began as an elite practice, apparently serving both as a means of sexual control (though the loss of mobility) and as an advertisement of status. It seems to have spread via an arms race: bound feet were a necessary condition of marrying up, and so, increasingly, all girls had their feet crushed, initially to improve their position in the marriage market, and then to have any position in the marriage market at all (Mackie 1996). As the practice becomes fixed, everyone is worse off than they were, but again no one can afford to defect individually (Edgerton 1992, 134–135). Some initiation rituals may invade by a dynamic akin to that of foot binding and female genital cutting. They begin as displays of courage, of the capacity to endure pain, and of endurance. They thus buy prestige and status. But then they become mandatory for all and hence leave their residue of costs without any benefits.

These commitment devices impose costs on the committing agent. But the costs are not Zahavian handicaps. A handicap signal and a commitment cost are crucially different. Commitment devices need not be *differentially costly*. Costly signaling is self-limited: a costly signal functioning as a handicap cannot spread through the population as a whole, for only the highest-quality individuals can afford to signal or can afford to signal at high intensities. Commitment signaling is not self-limiting in this way. The cost of the signal is an investment that changes the payoff matrix, reducing the signaling agent's temptation to defect. In principle, it can be in the interest of every agent to pay that cost. That is why female genital cutting, foot binding, and perhaps some initiation rituals can become universal practices, or near-universal practices, in a population. By supposedly reducing female sexual pleasure, female genital cutting reduces or eliminates the temptation to defect, by reducing defection's payoff. This mechanism does not depend on its being a minority strategy: all the women in the population can credibly reduce their temptation to defect by paying this cost. But once it is universal, none can *leverage* their position in the marriage market by committing. These women are caught in the fitness trap.[9] More benignly, it makes it possible for every individual in a group to establish trust.

## 5.7 Primitive Trust

The basic theme of this chapter is that cooperation—including earlier forms of cooperation that were the first foundations of human sociality—depends on trust. Trust is important directly in securing good partners for potentially profitable cooperative ventures. It is important indirectly in securing the profits of cooperation against free riders and bullies, for individual and collective threats of punishment, too, must be trusted. My problem, then, was to explain how trust is possible without presupposing the cognitive and social sophistication of late human sociality. My proposed solution depends on three ancient features of hominins and their lineage. One is the active role of hominins: we have long been niche constructors, acting on our environment, and as a consequence altering the costs and benefits of different actions at and across generations. Investment in relationships is a form of niche construction, one available to ancient hominins and one that can help explain trust.

A second strand—a recurring theme of this book—is coevolution. Suppose that different forms of cooperation began at roughly the same time—joint scavenging, mutual defense, reproductive cooperation (perhaps via crèching), mutual tolerance of first possession, antibully coalitions. Environmental change to increasingly arid and seasonal environments conjoined with very early technological innovation may well have been an important trigger for all these early forms of cooperation. If so, it would be no surprise if they first evolved at roughly the same time. If these do indeed appear together, each would help stabilize the other. The rewards of cooperation would sum over all these activities. The costs of exclusion and the benefits of reputation stabilize cooperation. There is a third mechanism: prosocial, affiliative emotions generated by joint success. Successful cooperation breeds trust. This internal, cognitive factor began, I suspect, as an exaptation of an ancient psychological mechanism. The amplifying effect of arousal was put to work in a new context; and while the great apes are much less cooperative than humans, some of those differences are limits in understanding rather than motivation. Tomasello and Warneken show that young chimps will help others without reward (when help is cheap) if they understand what is needed (Warneken et al. 2007; Warneken and Tomasello 2009; Warneken, forthcoming). So the evolution of prosocial emotions in hominins did not begin from a zero baseline.

In the wake of Robert Frank's brilliant and pathbreaking work on this issue, trustworthiness has often been modeled as an internal, subjective state of trustworthy agents, something they bring to interactions. Hence the problem of signaling that state in a deception-prone world becomes formidable. Commitment capacities must be shown through costly signals. I have argued that costly signaling models do not really fit commitment phenomena well, and have suggested an alternative conception of the problem. The subjective elements of commitment are often built interactively, in an interplay of joint action, shared experience, communication, and shared outcomes. Relevant changes in motivation and affect are built incrementally. Subjective commitment does not exist before and independently of others' recognition of commitment. It is built rather than displayed. Moreover, trust often depends not just on evaluations of the agent's subjective states but on the agent's public interventions. Trustworthy agents are trustworthy because they change their world to remove or reduce their temptation to break faith. The picture of trust is more interactive and interventionist, less a matter of signaling hidden preexisting psychological dispositions. There are multiple, sometimes mutually supporting, pathways to commitment and its recognition by others.

# 6 Signals, Cooperation, and Learning

## 6.1 Sperber's Dilemma

In chapters 1 and 2, I argued that cultural inheritance—the generation-by-generation accumulation of information—explains the striking contrast in social life and ecological style between hominins and chimps, and I introduced a basic model of that hominin cultural inheritance. In chapter 3, I applied these ideas to the human archaeological record, suggesting that they enable us to make sense of the otherwise puzzling transition to behavioral modernity, and that they offer a simple and parsimonious picture of the replacement of Neanderthals by *sapiens*, one that does not rely on undetectable, intrinsic cognitive differences between the species. Chapter 4 connected informational cooperation to other aspects of the human cooperation syndrome. Chapter 5 focused on the problem of free riding. Defection is controlled (though not eliminated) by the threat of exclusion from the benefits of cooperation and through direct sanction. But threats must be signaled, and this introduced the problems of honesty and trust. This chapter expands the discussion of signals, communication, and honesty.

In many species, agents learn from other adults about immediate threats and opportunities, as, for example, when birds in a flock respond to others' alarm calls. Indeed, social learning often does not depend on signals. One animal learns from another through information leakage. One bird in a flock spooks at a hawk overhead, and the surrounding birds respond to its reaction (Laland and Hoppitt 2003; Danchin and Luc-Alain 2004; Galef and Laland 2005). They benefit from one another's response, but that initial response is independent of the benefit it generates for others (this benefit of social life is known as by-product mutualism). So social learning is by no means rare, but its nature and role have been transformed in our lineage (Tomasello 1999b, 2008; Richerson and Boyd 2005). Much human cultural learning depends on signaling rather than exploiting information latent in the activities of social partners. Moreover, our cultural learning is often intergenerational: our signals transmit long-life information, often using arbitrary, low-cost signals (Laland and Hoppitt 2003; Danchin and Luc-Alain 2004). Moreover, human social learning is often transformative. A pied stilt that panics into the air as its fellows freak at a sudden sound will have essentially the same dispositions and capacities the next day. But humans acquire expertise, norms, habits, fears, and tastes that permanently change our phenotypes. As a consequence, as I have reiterated throughout the first four chapters, social learning is a dominating and pervasive feature of our life. We are habitual and obligate participants in rich networks of social or cultural learning. Much of the information (and misinformation) in my head is there because it was, earlier, in the heads of other agents. I am an information sink. I am also an information source: information (and misinformation) is in many other heads because it was first in mine.

This chapter explores the communicative capacities that made the central role of social learning in hominin life possible, and, especially, aims to explain the stability of information sharing in the face of potential defection.[1] The themes will be similar to those developed in the earlier chapters: stable cooperation depends on a set of coevolving mechanisms. No single key innovation explains honest and pervasive signaling in human life. We are, however, adapted not just to share information but to do so with some caution. The willingness to learn from others obviously has costs as well as benefits, and we are adapted to maximize the benefits while minimizing the costs. Dan Sperber has argued (in concert with others) that metarepresentation plays a central role in this optimizing process. Information sharing in a world of conflicting interest requires epistemic vigilance, and epistemic vigilance depends on metarepresentation. As with the Grandmother Hypothesis, I think that this idea contains an important insight, but in a modest rather than an extreme form. Epistemic vigilance is only one function of metarepresentation, and while the control of cheating in informational cooperation depends in part on metarepresentation, it depends on other mechanisms as well.

Cooperative information sharing is threatened by several different forms of defection. Collecting information is not free, so an agent could free-ride by not collecting information, thus sharing in the benefits of pooled information while paying none of the costs. A second form of social parasite would be an agent who exploited communal informational resources without being willing to share his or her distinctive expertise. These forms of defection compare to not hunting at all, and to hunting but not sharing, and so they are not distinctive to information-sharing forms of cooperation. The discussion in section 1.3 explains why the detection of free riding in small-scale social worlds is likely, and chapter 5 explains how such free riding can be suppressed despite the costs of control. Ecological parasites can be identified through leakage and through gossip and can be sufficiently well controlled by partner choice, incentives to invest in reputation, and the enforcement of norms. Similar mechanisms probably control informational parasites. Habitually failing to share information is as obvious in small-scale social worlds as other sharing failures. Everyone in a lab group or research team can identify those who are helpful and generous with their expertise. And they can identify those who keep their information to themselves. There is, however, a form of defection that is unique to information sharing and therefore poses a unique defection problem. Defection can occur via deception. Deceptive agents send signals that alter the behavior of the receiver in ways that have fitness costs to the receiver and benefits to the sender. The evolution of information sharing, especially sharing via arbitrary, low-cost signals, creates an opportunity to exploit others by means of false signals.

Thus epistemic vigilance is important, the idea goes, because communication using cheap and arbitrary signals is both risky and potentially profitable. Listening to another agent offers the opportunity to acquire crucial information at negligible cost. Such information can be of great value. Information about threats and opportunities can determine the course of one's life. The potential benefit of using others' information is extremely high; equally, the costs of trust can be catastrophic. In an environment of frequent informational cooperation and communication, the rewards are too great to be forgone, but the risks are too great to be ignored. Yet just in those cases where the benefits are greatest—where communication carries information about aspects of the world that are both important and expensive or impossible for the less informed to access—the veracity of the signal cannot directly be checked.

So we face a version of the last chapter's dilemma: the dilemma of trust. We cannot afford not to trust, and we cannot afford to trust the faithless.

So it is no surprise (Sperber argues) that we are unique both in our commitment to social learning and in our capacity for fluent metarepresentation. In an insightful series of papers, Sperber has connected these two features of human cognitive life, arguing that the evolution of fluent metarepresentation is largely a response to the increased importance of signaling—of cultural learning—in human life. While he finds no perfect solution to the problem of trust, Sperber suggests that folk epistemology—our set of tools for representing and evaluating thoughts, signs, and signals—is a partial solution. Metarepresentation evolves as part of a mechanism of indirect scrutiny, as part of a folk epistemology. We have a folk logic. We do not just represent representations; we assess them (Sperber 2000, 2001; Mercier and Sperber 2011; Sperber et al. 2010). Once we have a folk logic installed, we can assess messages for their coherence with what we know from other sources, and with what the agent has previously said. We can build epistemic profiles of agents: we assess the reliability of sources, as well as the plausibility of messages. These precautions are not perfect, and they are not free. But they are part of the trade-offs involved in trying to maximize the benefits of informational trafficking while minimizing the risks. In short, Sperber proposed an antiviral model of reasoning and of folk epistemology.

I think there is something importantly right about this idea, but Sperber's perspective on folk epistemology and its role in filtering deceptive signals is too Machiavellian. In this respect, the evolution of communication literature echoes the excessive focus on defection in the more general evolution of cooperation literature. The problem of trust is genuine and ancient: free riding and deceptive manipulation pose a risk to those engaging in information sharing, and that is true both of ancient and of contemporary social worlds. In addition, we face the more mundane risk of mere error: in a confusing world, honest intentions have never been a guarantee of a reliable signal. Obviously deception (and perhaps error and misunderstanding) has become a much more serious threat in contemporary mass social worlds, with their many one-off and arm's-length interactions. It is hard to think of a plausible Pleistocene analogue of Nigerian e-mail scams. But manipulation was a threat even in the small-scale, intimate social worlds of most human evolutionary history. But though real, this threat is not *uniform*. Other factors contribute to the robustness of honest signaling. The kind of information that flows, the nature of the channel, and the shape of the sender–receiver network are all relevant to honesty and deception. In sections 6.2 and 6.3, I argue that some forms of informational cooperation are much less prone to deceptive exploitation. In particular, I argue that the transmission of expertise is relatively immune to the problem of deception. That fact is important. It shows that some forms of informational cooperation can evolve and elaborate without requiring the prior or simultaneous evolution of complex cognitive tools. More risky forms of information sharing evolve after minds and environments become adapted to a complex and largely cooperative social life. I doubt that folk epistemology is as ancient as the dependence of hominin social lives on social learning.

Thus I argue that Sperber's dilemma is important but not ubiquitous. Moreover, policing defection is not the only role that folk epistemology plays in the evolutionary elaboration of information storing. I see reasoning and folk epistemology as multipurpose tools, playing broad roles in the organization and optimization of cultural learning. One purpose is to detect cheats, but another is to optimize the flow of honest communication. We can and do use broadly metarepresentational capacities to organize and improve the bandwidth and fidelity of information sharing, and, more generally, to act on our environment to improve cognitive efficiency. One distinctive feature of human intelligence is our capacity to organize our environment in ways that amplify the capacities of our wetware. Often these modifications depend on our capacity to respond to our own cognitive limitations,[2] as is shown in such ordinary (and probably ancient) acts as marking a trail in the bush to simplify navigation, making a distinctive scratch on an artifact to mark it as one's own, and using a tool as a template (which requires an agent to realize it can be used as an information store). There are many similarly low-key examples: I am monocular, and know it, and as a consequence I know that my judgment of distance, especially at night, is not reliable. I have to be more cautious than others in crossing roads at night. For the same reason, I have never acquired a driver's license. When this process becomes public and collective, reliability is much enhanced (for a recent review of the effect on reliability of collective decision making, see Conradt and List 2009). Andy Clark has devoted much of his career to documenting the amplification of human cognitive capacities with well-chosen external resources and demonstrating its importance (Clark 2008, 2010).[3]

So I see explicit argument and reasoning, and folk epistemology more generally, as having multiple functions. These include assessing the reliability of others' ideas, but they have many other functions as well.

## 6.2 Two Faces of Cultural Learning

In the previous section, I claimed that Sperber's trust dilemma was important but not ubiquitous. In this section and the next, I defend this claim and show its importance. The basic idea is that foul dealing poses a threat to some information-sharing transactions but not others because the opportunities, profits, and risks of deception vary. Before developing an analysis of the sources and consequences of this variation, I begin with a couple of illustrative examples. The first exemplifies the full-on Machiavellian dynamics that Sperber's analysis tracks. The second is a contrast case.

In the days before computer games took over, *Diplomacy* was a popular, though relationship-stressing, board game (I gather it has now morphed into an online game). The game is set in Europe, around 1900, with each player representing one of the European great powers of that period (counting the Ottoman Empire as part of Europe). The object of the game is to build a Europe-dominating empire through a judicious combination of alliance and betrayal. In *Diplomacy*, the paradigm communicative act is the conspiratorial whisper; the paradigm topic of conversation concerns future actions. The dynamic is Machiavellian. Agents are self-interested, but no triumph is possible without alliance, and no alliance is free of the risk of betrayal. Information sinks have no independent, direct test of signal veracity (until it is too late). But since blind trust is fatal, players must use imperfect, indirect tests. In particular, a sink must decide whether a source's stated intentions cohere with those that seem optimal for that source. Folk epistemology is a fallible tool, but it is the best one that agents have in managing and assessing conspiratorial whispers.

*Diplomacy* is indeed a model of one form of communication, and in such cases, folk epistemology does indeed play the role that Sperber identifies. But *Diplomacy* is not the only model of cultural learning. Consider a contrasting example from my youth: Monty Python's celebrated "Hungarian Phrase Book" sketch, in which a publisher produces a deceptive English–Hungarian–English phrase book in which, for example, the Hungarian phrase meaning "Can you direct me to the station?" is translated by the

English phrase "Please fondle my bum," and a protest about false arrest becomes "my nipples explode with delight." To those ignorant of English (or Hungarian), the adequacy of this translation is difficult to check directly (again, as in the first example, until it is too late). But there could rarely be a temptation to deceive in such a case. In part, this is because the phrase book is a public broadcast rather than a signal to a specific, pre-identified agent. As a result, the consequences of successful deception are much less easy to anticipate. For the same reason, successful deception is much harder to manage. Not all individuals who receive a widely broadcast signal will be ignorant in ways that make them vulnerable to manipulation. Their response can then cue those who are ignorant. The channel (language) is the same as that of *Diplomacy* conspiracies, but with the change in topic, and with the change to a multiagent, epistemically heterogeneous audience, the threat of deception essentially vanishes. Not all phrase books are well designed, but we discount the problem of deception for good reason.

These are toy examples, but they begin to reveal the complexity of information sharing and the variability of the threat of deception. In the discussion to come, I make four central points.

(i) Information sharing is indeed an instance of cooperation, and like other forms of cooperation, information sharing coexists with the possibility of free riding and deception. But this threat is contingent on the shape of the network that connects information sources and sinks, the communication channels through which information and misinformation flow, and the domain about which communication takes place.

(ii) In part because the threat of defection is so variable, folk epistemology is not just a policing mechanism. It is not just a filter that suppresses deception by making its detection more likely. It is also a set of tools we use to enhance the efficiency both of agent-to-agent information transfer and of individual exploration.

(iii) The evolution of cultural learning and information sharing is not a unitary phenomenon. Rather, it is a complex of coevolving but somewhat separate capacities. These include demonstration, gesture, and mime; language; joint attention and theory of mind; and observation learning. We have evolved the capacity to transmit and read signals sent through many channels. Those channels vary in their reliability, in their bandwidth, and in the kind of information that flows through them. Likewise we have also evolved ways of monitoring and intervening on those channels.

(iv) Finally, the evolution of the cultural learning complex is not just the evolution of individual adaptations to send, receive, and assess human signals. It is also the evolution of a distinctive form of social environment. In the first two chapters, I outlined and defended a model of human social learning, the apprentice learning model. I shall further argue that this form of social learning plays an important foundational role in the evolution of communication and information sharing because it does not generate free-rider temptations; there is little opportunity or temptation to deceive.

## 6.3   Honesty Mechanisms

*Diplomacy*-style cases are part of human communicative life. They are not artifacts of modernity. Such cases arise in the lives of foragers, and they do require the tools of epistemic vigilance that Sperber has identified. But they are not typical of human communicative life, and epistemic vigilance is not the only mechanism that keeps communication honest. In this section, I aim to isolate and discuss a set of honesty mechanisms. The overall aim is to show how an informational commons could evolve gradually without being subverted by epidemic deception. Reliable, low-risk information sharing can develop between agents without complex cognitive adaptations for cultural learning (such as metarepresentation-based vigilance mechanisms) even though such agents do not have identical interests. Once information flow across generations becomes typical and important, the selective regime changes. There is then selection for a set of adaptations that mediate social learning, including mechanisms that assess honesty and reliability. But our account of the genesis of information sharing must not assume that agents have such capacities early in this evolutionary trajectory.

I begin with the simple point that we have multiple channels through which information flows from one agent to another. These channels are not content neutral. It is, for example, much easier to send information about an animal by depiction (or mime) than by description. It is no accident that bird field guides have pictures and range maps as well as descriptions, and users rarely consult the descriptive text. It is even more difficult to convey useful information about birdsong by description than by imitation (many expert birders imitate calls remarkably well). So, for example, a standard field guide of Australian birds describes the call of the yellow spotted honeyeater as "loud, high pitched metallic, descending 'keah-keah-keah-keah!', machine gun rattle . . . loud scolding." Good luck, if you were using this description to distinguish this species from the very similar Lewin's honeyeater. Moreover, different channels expose agents to quite different risks of defection. I illustrate these issues in table 6.1.

I have organized this table so that early evolving, lower-risk channels appear at the top; later, often more flexible, but also more risky, channels appear at the bottom. I do not date the origins or amplification of these channels; there are great uncertainties in dating even fundamental hominin social and cultural transitions. Even so, to give some sense of the time depth here, it is useful to map this table onto Foley and Gamble's recent model of the major transitions in hominin evolution (Foley and Gamble 2009). Table 6.2 shows a summary version of their model.

Their "fourth transition" at about 400,000 to 300,000 years ago presupposes the evolution of high-risk signaling channels. Virtual communities, metacoalitions, and systematic and stable relations between communities all presuppose that communicative capacity has been decoupled from immediate context. Metacoalitions, for example, must organize and coordinate in advance, in anticipation of future events, rather than merely

**Table 6.1**

| Channel | Intrinsic reliability | Domain |
|---|---|---|
| Costly signals | High | First-party information, typically in bargaining and social competition |
| Information leakage via cues and economic activity | High (but not foolproof, as leaks can be faked or suppressed) | Skills, social information, local ecological information |
| Guided/structured trial and error | High | Skills, long-life information about the local environment |
| Imitation learning (eventually via demonstrations) | High | Skills, conventional gestures |
| Mimesis, gesture, depictive representation | Medium | Broad subject range, but perhaps not customs and norms |
| Language | Low | Very broad subject range, but often supplements rather than replaces other channels |

Table 6.2

| Transition | Time | Social implication | Structural social change |
|---|---|---|---|
| T1: bipedalism + range size expansion + fission–fusion | 6–4 mya | Communities more dispersed; more fission–fusion | Fission–fusion |
| T2: tools + meat | 2.6–1.6 mya | Enhanced male-female bonding | Male-female relationships |
| T3: fire + families | 800–700 kya | Further enhanced male-female relations (families) | Nested units within community |
| T4: social brains + spears + ecological expansion | 400–300 kya | Virtual communities; emergence of supracommunity structure | "Exploded" fission–fusion and metacoalitions |
| T5 ecological intensification | 200–10 kya | Elaborated inter-group networks; males control resources | Regional social systems; owned and inherited resources; depowering of women |

responding on the fly to contingencies. So if we accept anything like the Foley-Gamble model and its dates, hominins have used high-risk signaling channels adaptively for close to a half million years, perhaps much longer. These deep dates are reinforced by the morphological similarities between humans and Neanderthals. They were certainly physically capable of speech, with complex adaptations for breathe control and modulation of sound. That does not settle the question of Neanderthal language. Stephen Mithen points out that these capacities might be for song, not speech (Mithen 2005). But if Neanderthals could talk, our common ancestor probably had at least some capacities for language-like communication, and this too would suggest deep dates for sophisticated communication.

For my model, the absolute dates are less critical than the order in which these channels appear and become important. The apprenticeship model of cultural learning identifies a credible core around which an informational commons can accrete. In section 2.3, I claimed that one main advantage of the apprentice learning model is that it shows how expertise can flow reliably across the generations before the evolution of complex individual adaptations for cultural learning. It is surely likely that an agent can acquire the capacity to reliably make Oldowan-grade flakes and choppers on demand (and presumably some soft material technology—sharpened digging sticks and the like) in a hybrid learning environment without that agent having special adaptations for social learning. Such agents could learn by trial and error in an environment in which they would have many partial and complete templates to examine. Moreover, they would not have to find and identify suitable stone for themselves, and they would have many opportunities to supplement and guide their own trials by observing adult activity (for example, in trying to estimate the force with which they need to strike). If this model is right, skill learning can be important before the evolution of special social learning adaptations.

I can now supplement this argument by pointing to the fact that hybrid learning is a low-risk channel. In part, that is because it is hybrid. An agent acquiring a skill is interacting with, and experimenting on, the properties of material objects. These properties cannot be faked. The material world can be confusing, but it is not deceptive. But the social cues and prompts are low risk, too, for agents rely on information gleaned from the partial and complete products of skilled performance, and from information that leaks from the skilled in the course of their utilitarian, self-interested activities. Moreover, information is leaking in a partially or wholly vertical context. Early in these transitions to a world permeated by cultural learning, an agent's parents will be one source of information, even if they are not the only source.

Teaching—where an agent modifies his or her behavior at some cost, in ways that accelerate information acquisition by the sink—may not have been important until relatively late in this process, perhaps not until mode 3 technology or later.[4] But even when an information sink is relying on teaching—on the deliberately produced signals of the competent—this is a low-risk channel. A demonstration for teaching purposes is rarely identical to utilitarian performance. Demonstrations are slowed down and exaggerated; sometimes crucial elements are repeated. One point of demonstration is to make the constituent structure of a complex procedure obvious, for often that structure is not obvious in fluid, practiced performance. The skilled craftsperson need not operate according to a full-disclosure principle. She might keep special tricks to herself. But one with deceptive intent would face serious constraints, for despite the differences between demonstrations and for-real performances, the demonstration is constrained both by the material substrate and by the fact that it cannot be manifestly incongruent with utilitarian performance. An expert making fitted clothes could not, for example, fail to use a tool in demonstration that she manifestly uses in utilitarian production. Of course, deception remains possible. In showing how to detoxify a plant or prepare a hunting poison, a model might leave a crucial ingredient out or combine ingredients in the wrong proportions. But deception is constrained by the communication channel itself, not just the risk of consequences.

### Risk and Network Shape

The shape of the signaling network also reduces the risk of deception in skill transmission. Teaching and learning skill is often public. So, for example, the adze-making traditions described in section 2.3 involve skilled artisans working with their apprentices in a public space. Honesty is imposed on such transactions by the shape of the signaling network. Diplomacy depends on private talk; negotiations are one-to-one. In contrast, the Hungarian phrase book is public talk: books are sold to many users, and most users depend on many sources. Many-to-many networks are much safer than one-to-one exchanges. One reason why we need not fear fake Hungarian phrase books is that information about local conventions, customs, and norms is typically multiply and repeatedly sourced. Agents rarely learn to read the conventional, low-cost signals of their community—language, gesture, body language, local marks of status, role, affiliation, group identity— from a single individual, and still less frequently on a single occasion from a single other individual. In my politically depraved youth, when I was part of an ultra-left political sect, I learned the distinctive patois, gesture, attitudes, rituals, and public marks of my Trotskyist tribe by immersion, not by instruction from a single mentor. It would have taken a persistent and disciplined conspiracy (far beyond their meager organization talents) to practice deception upon me. We can rely on shared information about norms, customs, and symbols in part because there is rarely a temptation to defect. But redundancy is also important: we normally acquire information of this kind many times from many individuals. Would-be Machiavellis know this, and that constrains their deceptive options.

Network shape is important, but it is no silver bullet guaranteeing honesty, as both advertising and propaganda show. Even so, many-to-many networks enable an information sink to sample the opinions of many different sources, and that increases both the honesty and reliability of information flow, even when there is an initial information asymmetry in the network. Information pooling can be advantageous, even amongst equally well-informed agents, for it can enable each agent to increase the reliability of judgment in the face of environmental noise. This mechanism may be important in the evolution and stabilization of sharing information about rapidly changing features of the immediate environment, especially when those features do not have an obvious and unambiguous perceptual signature (for a survey of group decision making, and the role of information-pooling benefits in improving such decision making, see Conradt and Roper 2005). The Condorcet Jury Theorem makes the value of information pooling in the face of uncertainly vivid. If each juror votes independently and has a better than 0.5 chance of being right, as the size of the jury goes up, the probability of a majority vote being right rises rapidly to near certainty (List 2004). So agents gain access to reliable information about their environment if they have mutual knowledge of each agent's assessment of noisy signals, together with trust in consensus. Imagine a foraging party trying to decide whether a swollen river is too dangerous to ford, which animal in a pack to target, how to interpret the ambiguous behavior of a neighboring group. There is no temptation to defect here. By voting honestly and accepting consensus, each agent trades an unreliable assessment of a relevant feature of the world for a much more reliable assessment. Information pooling protects not just against deception but against noise. There is no temptation to withhold information, either. Doing so merely reduces the probability that the majority verdict is right.[5]

Notice that when agents pool information to increase reliability, the profit of cooperation does not derive from serial reciprocation; it is immediate. This shows that in some respects, information sharing is less subject to defection problems than some forms of ecological and reproductive cooperation. When cooperation has a physical product, conflict and defection can arise over fair division of the product. A jointly produced informational product—for example, a more reliable assessment of the risks of a river crossing—is automatically available to all who have pooled their individual estimates. In this respect, it is more like successful collective defense than successful collective hunting. Moreover, since information does not have the sensory salience of delicious food, or of an appealingly available but here's-trouble sexual partner, we do not seem to feel temptations to defect irrationally, preferring short-term reward over much greater long-term gains. As Richard Joyce (2006) has emphasized, our weak-willed, distractible, temptation-prone psychologies have doubtless often resulted in cooperation breakdowns that were bad for all involved. The human pleasure of gossip seems to suggest that if anything, we are tempted to share too much information rather than too little. Sharing information through gossiping and giving advice (whether wanted or not) does not go against our psychological grain; other forms of sharing quite often do.

I have just noted that socially acquired information about local norms, customs, and mores is more reliable because it is typically acquired in many-to-many networks. Network shape is also important to the evolution of a second form of information sharing: sharing ecological information that has a long shelf life. Hominin foragers range widely. Almost certainly, an expansion of range size drove the evolution of bipedal walking in our lineage (Foley 1995, 2002). It follows that in a heterogeneous and changing environment, no one individual is personally exposed to all he or she needs to know about resources and dangers, threats and opportunities. Unless early hominins foraged together in a single convoy, differing individuals and teams experienced differing spatiotemporal patches of their home range. Together with a fission–fusion social organization,[6] heterogeneity creates an information gradient (Dennett 1983, 1988) and thus a potentially advantageous division of epistemic labor. Everything has been seen by someone, but no one has seen everything; certainly no one has seen everything *recently*. In contrast to skill learning, the information channels on which ecological knowledge pooling depends are not intrinsically reliable. The flow of information depends largely on gesture, mime, and eventually language: low-cost signals about spatiotemporally displaced targets of inquiry (though no doubt some information flows inadvertently, via leakage).[7] Since the information flow concerns the elsewhere and the elsewhen, signal veracity cannot easily be checked directly against the world.

Despite the problem of low intrinsic reliability, the threat of deception is low. It is mitigated by two crucial—and, I suspect, stable and widespread—features of human social environments. The first we have already noted: information flow is often many-to-many. This makes both free riding and deceptive manipulation much harder. Those in the know will be aware of such defecting behavior, and the audience, too, will vary in the extent of their ignorance. What might deceive or manipulate one will not work on another. Public signaling reduces opportunities and temptations to defect.

Agents will often not be able to securely identify the information that they alone possess. If Abe, returning from a foraging trip, falsely reports that the figs on the other side of the river are not yet ripe, he had better be sure that no one else has been across the river or is about to go there.

There is a second mechanism: a gap between information pooling and information use. Structural information about the local habitat has a long useful life, so information can be and often is pooled *before* individual and collective deliberation and action. Agents are less tempted to hoard information because they will often not be able to assess the value of their private fraction of local knowledge. If people typically pump information into a common pot, they are less tempted to manipulate, because an agent who plants false trails will rarely know who will act on them or how. Even if an agent is prepared to deceive, he will often not be able to identify targets. He will not be able to estimate the effect of his deception on the world model of his victim, or how and when the victim will act on that world model. Many-to-many networks combined with sharing information in advance of action impose a veil of ignorance between a potentially Machiavellian agent and potential targets. That veil undermines the planning connection between false signal and Machiavellian consequences. So to the extent that local knowledge sharing is public and decoupled from immediate action, temptations to deceive are eroded. The risk of exposure is increased. The upshot, then, is that in public signaling contexts, the chance that an attempted manipulation will be detected is quite high. Its rewards will rarely be both high and certain. Since the individual and collective benefits of local knowledge pooling are significant, we can expect a default for honest signaling.

Sharing long-shelf-life ecological information was probably not one of the earliest forms of hominin information pooling to evolve. In contrast to skill transmission, this kind of sharing does seem to depend on capacities to send and interpret signals. But it need not depend on arbitrary symbols. As Kim Shaw-Williams has pointed out to me, gesture and other depictive formats have both the bandwidth and the specificity to encompass much ecological information. Animals, for example, can be identified by mime or by physically drawing their tracks. Location and distance, too, can be indicated by gesture, perhaps with some minimal and intuitive conventions that link signal intensity and repetition to distance.[8] The precision of these tools obviously has limits, but agents could share important ecological information without language or even protolanguage. Systems of gesture, mime, and depiction would suffice to share important environmental information in ways that will be kept fairly honest by public signaling and by pooling data before action. Relatively rudimentary signals can be both honest and cheap, and a form of information pooling can evolve that brings modest benefits to all parties. In small-scale, closed communities, the importance of reputation—the effect of your acts on third-party opinion—is likely to be a powerful mechanism in enforcing and reinforcing cooperation and deterring free riding, including informational free riding. So social information flow is important. But in contrast to ecological information, reputation-based enforcement of cooperation depends on complex communicative capacities. To gossip effectively, I have to be able to tell a story, to represent what others said and did at other times, places, and circumstances. Gossip seems to depend on something more like language.

## Risky Topics

The risk of deception depends in part on channel and network shape. But it also depends on topic. One important aspect of human social life is formal and informal bargaining. Obviously in bargaining interactions, the interests of the agents conflict. The more pigs a family extracts for their daughter, the less the groom's family can afford for their other sons. It is true that bargaining interactions often occur in the context of repeated interactions and an ongoing relationship, and this constrains the ruthlessness of the interaction and the degree to which it is wise to attempt outright lies. Even so, it will be in the interests of the family of the daughter to talk up her virtues, mention the interest in her from other quarters, and hint that the family is by no means desperate for her to marry at all right now. The groom's family, likewise, will need to exhibit a healthy skepticism about bridal claims and will meet bluff with bluff, exaggeration with exaggeration. Both sides will exploit independent sources of information to assess just how badly the other side wants to make the trade. And they will also bring their forensic skills to the interaction itself. Does the other side seem nervous or relaxed? Do their different claims cohere with each other and with background information? The less one has access to independent information, and the less the bargaining interaction takes place in the context of repeated business, the more wary a trader should be.

The size of the stake is important, too. *Diplomacy*-style bargaining is risky because the payoff from successful deception can be extremely high, and the sucker's payoff so utterly fatal. When payoffs are extremely high, reputation effects and costs of fracturing ongoing relationships no longer deter defection. Some high-stakes negotiation clearly depends on the institutions of modernity. Money, for example, makes it possible to control a valuable and extremely portable resource. Once money has been invented, it sometimes pays to cut and run. But high-stakes bargains were struck in small-scale traditional life, too. For example, Boehm's *Hierarchy in the Forest* (1999) shows that forager egalitarianism depends on the active defense of those norms and social practices. Would-be chiefs do arise in these societies and sometimes pose such severe threats that dangerous measures must be taken against them. Negotiating an enforcement coalition against a thug is a high-stakes activity, and often one that will need to be private. Likewise, in sexual negotiations the stakes are high. So high-stakes coordination and partnership decisions arise in small-scale societies, and with them comes a serious problem of trust. Moreover, trust requires more than confidence that one is not being lied to: joint plans miscarry catastrophically through failures of nerve, weakness of will, succumbing to temptation, and sheer incompetence. But while lies are not the only danger, agents do lie manipulatively in gossiping about others; they make promises and give guarantees that they never intend to keep. A liar induces others to act in ways that will benefit the liar but which have savage consequences for his or her targets of persuasion. So epistemic vigilance has sometimes been very important in traditional society.

Other domains come with much less temptation to defect. We acquire coordination tools through social learning. To become socially competent, children growing up in their culture (and outsiders who migrate in) need to absorb a lot of information about customs, norms, social signals, and rituals; taboos and the local targets of disgust; the local religion; the local language. For example, insiders recognize how sex, status, occupation, and social role are signaled by an agent's dress, bodily ornamentation, gesture, and distinctive forms of language. We acquire much of this information in many-to-many networks, and via information leakage from competent agents in utilitarian activities. As have seen, these are both honesty mechanisms. Moreover, we pick up much of this information—especially about gestures, customs of social distance, and similar norms of interaction—by unconscious processes of automatic imitation. Neither model nor mimic is aware that the mimic is noticing and matching patterns in the model's actions (see, e.g., Heyes 2011). But it is also true that there is little temptation to cheat. An agent who cannot read his or her social landscape will find it cripplingly difficult to engage in joint activities, cripplingly difficult to bargain and negotiate, cripplingly difficult to take advantage of what others know. Such an agent will have little value as a partner in cooperative enterprises. Indeed, in many circumstances, an agent who cannot read the social environment—who cannot communicate with, coordinate with, or influence others—is usually useless even as a target of exploitation (though occasional exceptions exist, for example, when agents who are socially clueless possess a valuable technical skill). The socially clueless will typically have no resources to exploit; they will be worthless as an ally, hopelessly unable to coordinate effectively in joint activity. Except in special circumstances, the socially clueless lack the capacity to help generate a joint product—a cooperation dividend—that the manipulator can then expropriate. No one gets rich defrauding the homeless. Only competent victims are worth exploiting; only the competent have something to steal. It is possible to generate fanciful scenarios in which one agent sets up a rival to be punished for norm violation by deceiving the rival about a local custom. But such scenarios are fanciful. In bargaining contexts there is a constant pressure to exaggerate or conceal. In contrast, one agent will rarely serve his own ends by withholding from another information about norms and meanings.

I summarize this section by returning to the examples of section 6.2. In both the board game and the phrase book, information flows via signals of low intrinsic reliability. But in other respects, the profiles could hardly be more different. *Diplomacy* involves one-to-one private signaling in high-stakes negotiations. The players cannot rely on the costs of relationship fracture and degraded reputation to prevent deception. Players exchange information about one another's intentions in the context of immediate plans for action: each agent can estimate how his or her information (or misinformation) will be used by the other. In contrast, information about English and Hungarian words is usually acquired in many-to-many interactions, increasing the risk that attempted deception will fail. Successful deception is difficult, often impossible, in many-to-many networks. But nor would it be tempting, for it is usually impossible to anticipate the contexts in which the misleading information will be used, and the disruption to social competence (even assuming a Machiavel could anticipate it) will usually just reduce the value of the stooge as a stooge.

## 6.4   The Folk as Educators

I began this chapter with Sperber's suggestion that fluent metarepresentation, especially as it is used in explicit reason and inference, is an epistemic filter. I end the chapter by returning to that idea to develop my suggestion that these capacities play a broad role in organizing the flow of information across and within generations. In section 2.3, I argued that amplified social learning and informational cooperation need not wait for special cognitive adaptations; social learning can expand through appropriate changes in ecology and social dynamics. But once hominins lived in social worlds in which cross-generational cultural learning played a crucial role in development, there was then selection for individual cognitive adaptations that improved the reliability and fidelity of learning. Researchers widely agree that our minds are adapted for social learning, and these adaptations enhance the reliability, bandwidth, and fidelity of social learning. It is true that individuals can learn skills by exploring their physical and biological habitat, and initial breakthroughs to new technologies, resources, and adaptive zones must originally have been made that way. But the crucial skills of *Homo sapiens* forager life could not be learned individually. One hundred thousand years ago, our ancestors had impressive technical skills. They had a full command of the ignition and control of fire. They used mode 4 technology in their stone toolmaking. This technology involved multiple stages of processing, as well as indirect force (striking out stone flakes by striking a stone on a hammer that was itself on a reworked core). It yielded sufficient control over form to make small, regular points. Ancient humans could attach these points firmly to shafts to make integrated composite weapons. Over the next 50,000 years there was more to come: more complex weapons (spear-throwers to go with their spears), tailored clothing, and a much more extensive use of other materials. But even the technology of 100,000 years ago was obviously far beyond the power of individuals to learn from scratch, even for a race of Faradays (McBrearty and Brooks 2000; Richerson and Boyd 2005). By then, and almost certainly long before, adult competence depended on minds adapted for social learning.

The nature of these adaptations remains a matter of profound debate. Despite a few skeptical voices (Dessalles 2007), surely language and the capacity to use it is one such adaptation (Tomasello 2008). One recent suggestion by Gergely Csibra and Gyorgy Gergely (with various coworkers) emphasizes the importance of pedagogy and hence converges with the themes of this work (Csibra and Gergely 2005, 2011; Gergely and Csibra 2005, 2006). The Csibra-Gergely model focuses wholly on the role of language, and while language is one crucial channel of cross-generational information flow, one of the main themes of this chapter is that other mechanisms are also important. But in other respects, their picture dovetails with the model of cultural learning defended here. First, the Csibra-Gergely model recognizes that both source and sink are adapted for the cultural transmission of information. Second, ecological and cultural information is learned simultaneously and in a mutually supporting way. Children learn novel artifacts or acts from demonstrations in combination with their linguistic labels and with the establishment of joint attention. Csibra and Gergely think that source and sink exchange cues that establish that the source will demonstrate new and relevant information, and these cues prime the sink's pedagogical interpretation of the source's acts. The capacity to learn from demonstration is part of a complex that includes sensitivity to cues of a teaching context (eye contact, motherese, imitation games, and the like), the capacity to interpret directional gestures as reference-fixing devices, and the expectation that the directional marker will be followed by new and relevant information about that referent.

Moreover, young children have the capacity to extract information rapidly in this context and detach it from its source. Instead of a child learning that Andrew likes broccoli,[9] she learns that broccoli is good (Gergely et al. 2007). In extracting information from these interactions, children act on the tacit assumption that information sources are benevolent, reliable, and typical. This assumption is adaptive, for as we have seen, this species of information flow—culturally specific information about objects, practices, and their names—rarely involves deceptive practices. Moreover, if the model is right, the pedagogy complex evolved and operated within family contexts that soften conflicts of evolutionary interest, and this too helps explain why a default assumption of honesty will not lead to exploitation and manipulation.

I do not know whether Csibra and Gergely's specific hypothesis is right; it is early days for their model. But I do think that fluent metarepresentation and folk epistemology more broadly understood are adaptations to cultural learning. We use our understanding of minds and representations to enhance the flow of expertise across the generations. One theme of this work and its ancestor (Sterelny 2003) is that humans are niche constructers par excellence; in particular, we organize, adapt, and enhance the learning environment of the next generation, both individually and collectively. We construct the epistemic niche of the next generation. The capacity to represent our own cognitive, communicative, and learning capacities is not essential to epistemic niche construction. Good tricks can be discovered, recognized, and retained by trial and error. But the ability to represent our own cognitive capacities certainly enhances epistemic niche construction. Consider, for example, teaching. Effective teaching by demonstration requires an agent to make the structure of his or her skill overt, and doing so requires models to represent their own capabilities. Expert performance is often rapid and fluent, without obvious components. Learning from such performance is difficult. It becomes much easier if the task is overtly decomposed into segments, each of which can be represented and practiced individually. So fluent natural performance is often less useful as a model than performances that are stylized and accompanied by metacommentary. Such a stylized performance, with its commentary, requires models to represent to themselves their own competence.

More generally, the active supervision of learning requires experts to understand what the inexpert can and cannot do, so that experts can assign tasks (and suggest exemplars) that lie within the nonexpert's capacities but stretch or consolidate them. Assigning tasks, providing exemplars and examples, and setting learning goals so that each task prepares for the next all improve the reliability and fidelity of learning. In short, agents exploit the resources of folk epistemology when they think about how to make what they know easier for others to learn, and when they adapt their own activities to compensate for their less-skilled juvenile dependents. Organizing learning requires an agent to understand a skill, not just possess it. Of course, experts are not on their own here. Just as expertise is a multigenerational invention, so lore will often be available about the best practices for the transmission of expertise, lore that is itself a multigenerational construction and is itself an aspect of folk epistemology. Folk epistemology is important for the sink, too. In some circumstances, sinks need to choose their source, especially as they begin to explore their environment away from their immediate family. They need to be able to identify expertise.[10] They need to understand that a performance is stylized, that a crucial step has been slowed down, exaggerated, or repeated to make it more overt. In short, the sink and the source need to read each other. Each monitors the other and their joint focus of attention and intention.

If folk epistemology is a learning and teaching tool, as well as antiviral software, then we should see a correlation between the complexity of a skill and the capacity to represent one's own expertise. More complex technologies should map onto increases in an agent's capacity to represent the structure of his or her expertise. The high-fidelity transmission of complex technique requires experts not just to organize their skill as a behavioral program but to be aware of that program and to be able to action elements of it independently. Likewise, complex technologies should map onto an expert's capacity to choose helpful examples and sequence learning tasks optimally so that the inexpert acquire and practice subskills in the right order. Equally, as skills and technologies become complex and teaching becomes important, we should see an improvement in diagnostic skills, an improvement in an expert's capacity to identify just what is wrong with another's performance, not just to realize that something is wrong. In short, if folk epistemology plays an increasing role in making cultural transmission more reliable as skills become more demanding, we should see a correlation between an increased complexity of expertise and self-reflective expertise.

I finish with a much more speculative thought, connecting the pace of development to the demands of social learning. Metarepresentational capacities develop rapidly. They are important in the lives of young children, children as young as three or four, perhaps younger. Such children are not economic agents; they have no need for these capacities to coordinate collaborative resource gathering. Nor are they given to playing social or sexual chess, jockeying for leverage in a competitive world. Their fitness interests are probably not much threatened by deceptive practices of siblings and contemporaries. But they do face a pressing problem: social learning. If the model of life history evolution discussed in chapter 4 is right, the shape of human life history in general, and in particular the length of childhood and adolescence, is explained by the enormous size of this task. So, perhaps, is the early onset of apparently sophisticated metarepresentational capacities.

Michael Tomasello and his colleagues have argued that joint attention, and then joint intention, emerge very early in the life of a child.[11] They argue that joint attention, joint intentionality, and human cooperation are intimately related (these ideas are well summarized in Tomasello 2009). Joint intention is not characterized precisely, but it involves a complex of capabilities that include reading another's thoughts while being aware that yours are also being read, together with motivational and representational elements. If an action is truly collective, part of the reason for engaging in it is that your social partners are engaging in it too. These motivational and representational aspects are important; joint intentionality comes with a particular cognitive capacity: the capacity to decompose complex joint activities into "agent-neutral" subroles. We represent the division of labor required to execute a collective goal in ways that do not tie specific roles to specific agents. We organize collective action through representing multiplayer behavioral programs. These elements do not come onstream simultaneously. However, Tomasello and his colleagues argue that this cognitive complex develops early, with some aspects of joint attention in action around one year of age, and important cognitive and motivational elements in place by three.

In a parallel development, the last few years have seen a reevaluation of accepted wisdom about understanding false belief. The standard line has it that children do not understand false belief until around four; at that age, but not earlier, they understand that other agents will act on their own representation of the world, even when it is false. The revisionary line suggests that these classic experiments overloaded young children, and so a new generation of false-belief tasks depend on data about young children's direction of gaze (and similar behavioral cues); the new tests do not rely on asking three- and four-year-olds what they expect to happen (for a review, see Baillargeon, Scott, and He 2010). I am not yet convinced. We should be cautious about interpreting young children as having rich and explicit metarepresentational capacities, including the capacity to attribute shared or false beliefs when they cannot manifest that capacity in language, despite having apparently mastered the necessary linguistic tools. But even in the standard picture of the development of mind reading, these abilities emerge early. This early emergence offers some support for the idea that the function of mind reading skills is to enhance the flow of social learning early in life, not just to filter and protect somewhat later in life.

This chapter has focused on the mechanisms that maintain the complex web of information sharing on which behaviorally modern human life depends. The overall picture has six core features. First, it identifies a low-cost, low-risk starting point that can midwife the evolution of more complex, special-purpose adaptations. Second, it emphasizes the role of the material substrates of learning and teaching. Social learning is usually hybrid learning, and the physical features of artifacts and raw materials channel and support learning. Third, it identifies an important class of social circumstances where free riding is difficult or impossible; these build foundations for the extension of communication into more risky contexts. Fourth, it highlights the differences between signaling channels; these differ both in the kind of information that flows through them and in their reliability. Fifth, in contrast to much formal modeling, the picture does not take one-to-one private talk to be the base case. Public information pooling is, I suspect, early and important. Finally, here and elsewhere, the model is coevolutionary. It identifies no single mechanism or adaptive breakthrough on which human social learning rests.

My aim has been to explain why our rich, complex forms of information sharing remain stable and powerful despite failures to contribute, attempts to deceive, and simple unreliability, and to explain the evolutionary construction of this complex. How did our lineage shift from the limited version of information sharing that characterizes great ape life to the complex, stabilized, and obligate information sharing of behavioral modernity? I have identified forms of information sharing that are plausible candidates as early evolvers. They are profitable, and that profit does not depend on the prior possession of specific cognitive adaptations for cultural learning or information sharing. They are low-risk forms of information sharing. Skill transmission, fast-fading ecological information, and long-shelf-life ecological information are all likely candidates for being early evolvers. The establishment of these early evolving forms of informational cooperation in hominin social worlds changed the nature of those social worlds, leading to selection for individual cognitive adaptations. Those extended the range of cultural learning, improved its fidelity and bandwidth, and managed the risks of deception. Moreover, the interaction of informational and ecological cooperation increased information gradients in hominin social groups and thus increased the potential profits of information sharing and communication. As forager skill levels increased, so did life expectancy, intensifying an intergenerational information gradient. Specialization and the division of labor likewise contribute to informational gradients, for specialization results in agents exploring different aspects and areas of their common range. These steeper gradients amplify the potential profits of communication, thus contributing to the positive feedback between cooperation, communication, and technical intelligence. The opportunities for mutually profitable informational cooperation increased as social life elaborated. Human communication is not always frank and honest. We lie and conceal. But human communication is typically cooperative: we do not always have to filch information from others or wait for their expertise to be manifest in their workaday lives.

# 7 From Skills to Norms

## 7.1 Norms and Communities

The theme of my whole argument is that humans are unique largely because we have evolved the capacities to accumulate and use cognitive capital. I have suggested that information pooling at and across generations has long been a central feature of human life, but the nature of this pooling has been transformed over time. I suggested in sections 2.3 and 3.4 that the power and fidelity of cross-generational flow increased by the increasing organization of the learning environment, by the gradual evolution of specific adaptations for learning and teaching, and by the invention of cognitive technologies. Language, most obviously, is an immensely powerful tool for both learning and teaching. These adaptations increase the capacity of existing channels, for example, as adults shift from merely tolerating observation learning to actively demonstrating crucial components of skill. And they add new channels.

One message of the last chapter is that there has also been a revolution in the kind of information that flows from one generation to the next. The social acquisition of expertise about the local environment and the resources it offered evolved early in the hominin information-sharing career. These continue to be crucial: the profit of cooperative foraging continues to depend on coordination, skill, and expertise about the local environment. But as noted in sections 3.2 and 3.3, human social life transformed as we evolved from simply living in groups to being aware of, and identifying with, the groups in which we live. In the terminology of archaeology, we came to live in symbolically marked groups, consciously identifying as members of specific groups and manifesting that identification by distinctive language, customs, style of dress, ornamentation, and the like.

Symbolic marking tends to be stable across generations, so this change is associated with, and dependent on, the social transmission of new kinds of information: information about norms, customs, and symbols. The shift, I presume, was gradual. But children came to learn not just the tools of their ecological trade and the characteristics of their local world. They had to acquire the distinctive language, mores, customs, attitudes, beliefs, and public symbols of their group.

These may be particularly difficult learning targets. Norms and attitudes are not observable in any straightforward sense. Customs, and especially norms, are more than mere regularities in appearance or behavior. Mere observation of agents in action will not distinguish does-not-happen from must-not-happen. Sometimes they are not even behavioral regularities; norms can be endorsed, accepted, but not always observed. By the time that behaviorally modern humans evolved (perhaps much earlier), children faced these new learning demands. According to one influential line of argument, children learn the norms of their local culture through cognitive adaptations specific to this very learning task. Children acquire the norms of their local world, and in particular the moral norms of their social world, because they have crucial information preinstalled. We are innately adapted to learn morality, perhaps even of a highly constrained kind. As with other nativist hypotheses, the extent to which human minds are supposedly primed to learn moral norms varies greatly from theorist to theorist. According to one line of thought, we have a norm acquisition device, but not one that determines that we acquire specifically moral norms (Sripada and Stich 2006). Others believe that no normative principles are innate, but core moral concepts are (Joyce 2006). But the best-known and most widely discussed version of moral nativism explicitly treats language as a model for moral cognition, and that will be my foil in this chapter.

Nativist models of norm acquisition understate the variety and novelty of the normative environment to which humans adapt, and they understate the power of hybrid learning strategies in engineered learning environments. Thus I argue in this chapter that the basic machinery introduced in section 2.3 and elaborated in later chapters suffices for this learning challenge. Explaining norm acquisition is important to my project. The normative life of a community is a distinctive feature of human life, so it is important that the feedback-driven model of evolution defended here can explain the establishment and transmission of norms. But it is also important as an exemplar of extended hybrid learning. In section 2.3, I briefly drew on ethnography to show that intellectual skills are sometimes transmitted in apprentice-like learning niches. I think the same is true of the flow of norms across the generations: children learn norms reliably by using hybrid learning strategies in an organized learning environment. Hybrid learning has been exapted for the transmission of local mores, customs, and norms. In this chapter, I begin with a brief account of the nativist picture and then explain my skepticism. However, I agree that moral learning is biologically prepared, so I offer an alternative model that depends both on perceptual biases that make the emotional effects of action salient to us, and on informationally engineered developmental environments. As in other critical cases, the flow of information across generations depends on an interaction between organized learning environments and individual cognitive adaptations. The acquisition of moral competence depends on mechanisms that are similar to, and are modified from, those that explain skill acquisition.

## 7.2   Moral Nativism

Marc Hauser (and his colleagues), John Mikhail, and Susan Dwyer have recently developed an extended parallel between moral and linguistic cognition, arguing that we have a "moral grammar."[1] The parallels they draw between normative and linguistic cognition are important, because language is the least controversial example of a special-purpose human cognitive adaptation.[2] Language is independent of conscious cognitive processing. An agent does not have to decide to hear speech as language; it is automatic, mandatory. Agents parse sentences of their native language, recovering their organization. But they have no introspective access to the mechanisms that take speech as input and deliver to conscious awareness an interpretation of what has been said. They recognize in aberrant cases that something is defective about sentence organizations. But agents do not have introspective access to the information about their native language that they use in such judgments. This information is tacit. It is not portable; that is, it is not available to drive other cognitive processes and hence cannot be expressed as assertions or beliefs. Moreover, the acquisition of language is rapid, with a predictable sequence, and arguably from an informationally impoverished experiential input. So the normal development

of the human capacity to use language may result from an innately pro-
grammed language module coming onstream. The human mind comes
pre-equipped for the task of learning and using language (the most vivid
exposition of this idea is Pinker 1994, though Noam Chomsky is its source
and standard-bearer).

Hauser, Dwyer, and Mikhail suggest that the human capacity for moral
judgment has a structure parallel to that for language, both in development
and operation. The moral nativists argue that moral cognition is not just
universal but universal in surprising and subtle ways. Thus Hauser (2006)
argues that there is striking (though still only suggestive) evidence of cross-
cultural uniformity in moral judgments. Most agents draw an important
moral distinction between acts and omissions (though not all agents; indeed, one recent report examines a culture that systematically rejects the
moral significance of this distinction [Abarbanell and Hauser 2010]), and
they draw an important moral distinction between the foreseen but unintended consequences of actions and those consequences that are both foreseen and intended. So in pursuit of the best overall outcome (for example,
saving the most lives possible in the face of imminent disaster), it is morally
acceptable to tolerate bad consequences that are foreseen but unintended.
But it is not morally acceptable to intend bad consequences, to intention-
ally do evil to block a still greater evil. So they judge in accordance with the
so-called principle of double effect. In developing the grammatical model
of morality, Hauser depends on the subtlety and the relative cultural invariance of these general moral principles. But he also emphasizes the introspective opacity of the principles that appear to guide specific judgments.
Although the principle of double effect seems to shape moral judgments,
without prompting, few people can explicitly state that principle. So it is
central to the nativist case that moral judgments systematically depend on
tacit principles. We can predict agents' moral views. They reliably judge
that some actions are morally appropriate and others are not, but they
can rarely articulate the principles that guide such discriminations (Hauser,
Young, and Cushman, forthcoming a).

In part, Hauser bases these claims on online survey work: subjects logged
on to make judgments about moral dilemmas based on a famous class of
thought experiment involving a streetcar or trolley running out of control,
putting bystanders in mortal peril. Subjects read about scenarios in which
an agent can choose to save five lives, but at the cost of one. The cases

differ in the kind of interventions that are necessary to save the five. In some interventions, the agent must directly kill the one (using him as a brake, buffer, or obstruction). His death is the direct target of intervention; it is the means of saving the others. For example, to save the five from a runaway trolley, an innocent bystander must be pushed onto the tracks to slow the trolley, giving the five time to escape. In others, the one dies, but as a side effect. The intervention to save the five does not in itself depend on its being fatal to the one. In one scenario, the agent has access to a switching mechanism that can divert the trolley from one line to another. Unfortunately there is an innocent bystander on the spare line, too, but though his death is foreseen, it is not intended. The five would be saved even if he were absent.

Many people confronted with these cases consistently discriminate between the two types of intervention. They judge that it is permissible to sacrifice the one as an unfortunate side effect of saving the many, but it is not permissible to deliberately kill the one as the direct means of saving the many. However, few can specify the principle that divides the cases into two kinds. Their moral judgments are robust and replicable but depend on principles that are introspectively inaccessible. If moral judgments depend on principles that are rarely articulated, it is hard to see how an agent could learn those principles, especially if they are not obvious generalizations from particular cases. So if Hauser's picture is right, moral development confronts a poverty-of-the-stimulus problem. We must be prewired for specific moral principles, for we come to rely on those principles in our fast and unreflective moral reasoning although they have never been articulated and taught in the period in which children become moral thinkers. The moral nativists agree that moral development involves significant cultural input, but they also argue that it has a significant and specifically moral innate component.

## 7.3 Self-Control, Vigilance, and Persuasion

If the moral grammarian's model applied seamlessly to our moral cognition, we would find ourselves with strong moral responses to specific situations, but with neither the capacity nor the inclination to offer any explanation or defense of those responses. The double-effect scenarios result in consistent judgments, but stumbling and inarticulate justifications of those judgments. If moral cognition really did depend on something organizationally like a grammar, that would be typical. But it is the exception, not the rule. For example, another factor that seems to make a difference to intuitive judgment is whether the focal agent is acting or merely refraining from action. Suppose there is only a single line along which a runaway trolley is hurtling toward five trapped individuals. Most think it wrong to tip an onlooker onto the tracks, but permissible to refrain from lowering down a rope ladder, enabling an isolated person in front of the five to escape, but dooming the others. Acts and omissions are not judged in the same way, and many agents can articulate a distinction between act and omission. Moreover, we are rarely indifferent to the challenge to articulate and defend the principles behind our judgments. As we shall see shortly, these facts do not fit the grammar model of moral cognition well.

If that model were right, moral judgments would be more like disgust responses. Quite often, the foods of one culture will strike the members of another culture as simply disgusting, but those who respond with disgust are unable to articulate any clear basis for their disgust. Many Anglo-Irish Australians think it is disgusting to eat dog, or even horse. As the nativists emphasize, we do have similarly fast and reflexive moral responses, and sometimes we cannot articulate their rationale. But that is not all we have: we can and do articulate moral thoughts, even if it turns out that there is a mismatch between our articulated and our reflexive responses. The moral grammarians are not well placed to explain our capacity to reflect on and articulate moral judgment, for they model moral cognition on an aspect of language where we have almost no capacity to articulate the principles we use. We do not have introspective access to grammar, to the syntactic principles that we use in organizing words and phrases into sentences and use in decoding the speech of others. We have a folk semantics: we are able to talk about meaning, truth, possibility, necessity, impossibility. But we have virtually no folk syntax. Unless syntactic categories are explicitly taught in formal education (verb, noun phrase, complement, etc.), children do not acquire them as a standard by-product of language acquisition and enculturation. The same is not true of, say, "false," "likely," or "impossible."[3]

The syntactic model of moral cognition has no natural model of the relationship between reflexive and reflective moral cognition. Suppose Hauser and his colleagues are right to think that an abstract set of normative general principles develops in every normal individual, principles whose general character is invariant but whose specific form depends on individual learning history, principles that automatically, rapidly, productively, and unconsciously generate moral appraisals. Even if all of this is true, language and moral cognition differ in striking ways. For in contrast to language, every normal agent also has conscious moral principles. We endorse moral generalizations; we do not just make bullet-fast judgments about specific cases. What is the role of reflective morality in the Hauser-Mikhail-Dwyer picture?

One idea is that reflective morality is disconnected from agents' actual practices or moral evaluations. We make judgments and endorse moral principles, but the stories we tell ourselves and others have no systematic explanatory connection to the judgments we make. The features that we say drive our judgment do not actually do so. Reflection is confabulation, and our reflective moral principles play no role in our day-to-day moral practices. They are epiphenomenal. A moderate version of this idea is quite plausible. Jon Haidt and his colleagues have argued that we overstate the importance of reflective morality and explicit moral deliberation. Often moral reflection is the post hoc rationalization of rapid, emotionally mediated responses. These authors point to striking experimental data showing that moral judgment is significantly influenced by irrelevant features of context. For instance, if experimental subjects are placed in a foul environment (dirty ashtrays, litter around) and asked to judge scenarios involving potential moral violation, they judge much more harshly than subjects who evaluate similar scenarios while in a pristine environment. But Haidt and his colleagues certainly do not argue that conscious moral reasoning is epiphenomenal, especially not when it is collective.[4] Their caution is surely appropriate. No doubt conscious moral reasoning is sometimes confabulation. But it would be implausible to claim that reflective morality is wholly epiphenomenal, for agents seem to change moral practices (converting to vegetarianism, for instance) as a result of newly acquired reflective moralities. Over time, reflection influences reaction.

Another thought is that reflective morality is incomplete introspective access to the moral principles that drive our moral module. Experimental work seems to show that agents have some conscious access to the principles that underlie their judgments. As we have seen, subjects can articulate a principle appealing to the moral difference between acts and omissions (Cushman, Young, and Hauser 2006). Judgment also seems sensitive to the directness, the causal intimacy, of an agent's intervention, and subjects are often aware of this factor. But moral vegetarianism and similar cases trouble this idea, too. Such examples show that reflective morality is unstable in individual agents and variable across agents. Agents convert to utilitarianism, rejecting the distinction between act and omission and the doctrine of double effect. In contrast, once triggered, with its parameters set, a moral grammar is presumably fixed. Moreover, agents can have, and be aware of, a tension between their reflectively endorsed moral views and their reactive moral response. So the moral grammarians seem committed to a dual-process model of the mind (e.g., Stanovich 2004). We have both tacit and explicit moral cognition, and the systems interact. Syntax can offer no model of this interaction, because syntactic processing does not depend on conscious reasoning at all. So the moral grammarians have no model of the interaction between the fast, tacit, automatic system and the slow conscious system, as Dupoux and Jacob also point out, though without convincing their targets (Dupoux and Jacob 2007; Dwyer and Hauser 2008).

So language processing and moral cognition differ in that we have reflective, conscious, general moral principles that interact with fast moral response; but we have almost no reflective, conscious, general principles of syntax, let alone ones that make a difference to language processing. There is a second crucial difference. General principles of language are opaque to introspection, but so are structural representations of *particular utterances*. When we hear and understand a sentence, in some sense we must represent its organization or structure, for that organization plays a central role in its meaning. We understand sentences, and to understand them, we must represent sentence structure. But few of us have syntactic beliefs about specific sentences. The parsing system takes as input sound from speech and gives as output a representation of structure. But that output does not consist of beliefs.[5] In contrast, the output of a normative judgment system (if there is one) consists of beliefs about what should or should not be done. We may not always be able to articulate *the general principles* that guide our normative judgments. But we can certainly articulate those judgments themselves and make them public. We do not just find ourselves inclined to act one way rather than another. Moreover, these judgments are thick rather than thin. Our culture provides us with a rich and nuanced moral vocabulary. So, for example, "cruel," "capricious," "spiteful," "kind," "vengeful," and so on are all terms of folk morality: they are terms of normative evaluation available to ordinary folk. Law and moral philosophy have developed technical vocabularies, but these vocabularies are built on, and continuous with, the rich resources available to nonspecialist agents. That is not true of the specialist vocabularies of linguistics, psychology, or economics.

That is no accident. Richard Joyce, Robert Frank, and others argue that moral cognition is an aid to individual self-control; moral thought inhibits succumbing to temptation (Frank 1988; Joyce 2005). Joyce, for example, suggests that we are apt to discount future benefits, and that this undermines our capacity to cooperate through prudential calculation, although such cooperation is almost always prudent. Perhaps enhancing self-control is one function of moral cognition. If it were the only function, a disgust-like, purely reflexive mode of moral cognition would suffice. But it is not. Moral judgment is also a tool of social influence. We seek to persuade others to share our moral views both in particular cases and with respect to general principles. Haidt and Bjorklund (2008) quite rightly emphasize the fact that moral reasoning is often collective and social; it is a central part of social life. Moralizing is not a private vice. In making normative appraisals, agents are not just in the business of guiding their own behavior; they are in the business of guiding and evaluating the actions of others. This is a core feature of moral cognition, and one that contrasts with syntactic judgment. Moral persuasion has no syntactic analogue. In general, the more invisible syntactic processing is to the users of language, the better. It is like our retinal image. Artists excepted, we do not want to know about our retinal images, or anyone else's. Rather, we want to, and we normally do, "see through" the image to the world; we want to know what others see, not what they see with. But we have excellent reasons to identity the normative judgments of others. They are important indicators of how those others will act. And when we disagree, those views are the targets of our persuasive efforts. This public, persuasive role requires reflection, making it necessary to articulate the bases of intuitive judgments. If the moral assessment of the acts of others was nothing but policing unambiguous norm violations, once again an unreflective, disgustlike mode of moral cognition might suffice. But I doubt whether human moral worlds have ever been unambiguous. Even in small, closed communities with agreement about general principles, the application of principles to particular cases would always have been contentious. So persuasion and hence norm articulation must always have been important. The role of moral judgment in human social life requires specific moral judgment (and perhaps the principles that guide those judgments) to be in the public domain. It is no accident that we engage in explicit moral thought.

## 7.4   Reactive and Reflective Moral Response

Much moral assessment is fast and automatic. We do not have to decide whether to evaluate a situation normatively when we read of the abduction of a three-year-old girl. Moreover, there does seem reason to suspect that normative thought is a universal feature of human society and of normal agency. The capacity for moral cognition seems to develop impressively early and with impressive robustness. Consider, in contrast, literacy and numeracy. Their acquisition requires extensive formal education, but despite that formal education, the development of these skills can be fragile. Phenomenologically, moral cognition does share some of the characteristics of the classic innate modules: the perceptual systems and language. Those who are skeptical of the moral grammarians' picture will need to explain why moral cognition seems modular, if it is not.

It is much less clear, though, that moral thought shares the cross-cultural invariance of, say, vision. Indeed, normative judgment could not be invariant in the same way that the recognition of facial emotions or color discrimination seems to be. The moral norms expressed by a culture are intimately linked to that culture's social structure, and structure varies remarkably across time and space. Social systems vary in their horizontal and vertical complexity; the organization of gender relations and family structure; their economic basis; the extent of social, economic, and political inequality; population size and connectivity; and religious and ritual commitments. Moral norms regulate action in all these highly variable aspects of human social worlds, so it is no surprise that we see profound normative differences across cultures. Perhaps some of these differences are not differences in moral norms, properly understood. Liberal democracies now distinguish between genuine moral norms and those of religion and disgust, so some of this normative variation might be soaked up in these other categories. That may be so, but the distinction between such norms is less clear in other cultures (Machery and Mallon 2011). In any case, our explanation of normative cognition must include this whole package, one that clearly varies significantly from culture to culture, even if there are also common elements and family resemblances.

Variation is not just variation in reflective morality. The recent history of the Western world shows massive shifts in intuitive response, not just reflective morality. I grew up in an Australia in which homosexual acts caused moral disgust. Fortunately that has changed. This is one of many examples. In recent years there has been a shift in intuitive response to animal cruelty, the appropriate discipline of children, the role of women in the world, deference to social superiors, and to being overweight. History shows considerable change in our visceral response to many situations. Not so long ago, public execution was an immensely popular public entertainment in England. History and anthropology give us good reason to suppose that reflective and intuitive morality influence one another. They interact through life, not just in a critical period of childhood or early adolescence, when the child is learning the specific local form of the universal system of moral principles. It by no means follows from this that literally any system of imaginable normative beliefs is a humanly possible moral system. But there is no reason to believe that the spectrum of possible moralities is narrow, and so cross-cultural work is important to determine whether Hauser's initial results are robust. These results give us reason to suspect there are family resemblances between the moral systems of different cultures.

The phenomenology of moral response—its rapidity and insensitivity to top-down control—is important. We cannot decide not to respond to perceived cruelty with outrage, though we can inhibit action. But I do not take that to be a persuasive consideration favoring nativism, for expert judgment is fluent, unreflective, and automatic. An expert birder, for example, cannot help seeing a red-rump parrot as a red-rump, however much he would prefer to see it as the similar but much rarer orange-bellied parrot. Natural minds are good at learning to recognize patterns—similarities between instances—and the exercise of that capacity results in intuitive judgments about new cases. Pattern recognition is productive: it extends to indefinitely many new cases. Such judgments are fast and automatic, and often the agent cannot explain the basis of the judgment.

In a series of vivid examples, Chomsky argued that grammar could *not* be captured as a set of simple lexical patterns. We can capture the relationship between (say) indicative and yes–no questions only by identifying the abstract structure of a sentence, its organization of lexical items into sub-sentential constituents. Chomsky's argument is crucial. If sound, it shows that syntax is not pattern recognition. Agents do not learn their language by generalizing from familiar cases, using familiar examples as models for new constructions (though see Tomasello 2003 for a view of language acquisition appealing to such mechanisms). No Chomsky-style argument has been run for moral cognition. While Hauser has certainly argued that even elementary moral judgments depend on subtle features of situations, those features are not moral but intentional. They are facts about an agent's goals, intentions, and consequences (see Hauser 2006; Mikhail 2007; Hauser, Young, and Cushman 2008a,b).

So my working hypothesis is that our intuitive moral judgments are generalizations from exemplars. Kind and generous acts are those that resemble paradigmatic moments of kindness or generosity, and so on for other evaluations. Pattern recognition is fast and automatic once the abilities have come online. Moreover, pattern recognition can be tuned to abstract rather than sensory properties. Chess experts, for example, assess positions rapidly and accurately, and a chess player has great difficulty not seeing a chess position as a position. An expert cannot help seeing a knight on E6 as a knight dominating the center, even if he can also see it as a shape on a pattern of squares. When a master recognizes a position as one that favors a knight over a bishop, the relevant properties are subtle, geometrical features of pawn configurations: patterns that determine the relative mobility of the two pieces. Moreover, the features and weights that underlie pattern recognition are often tacit. Expert birders, for example, can often recognize a bird from a fleeting glimpse without being able to say how they recognize it. They will say that it has the "jizz" of a brown falcon. So if intuitive moral judgments are the result of capacities for pattern recognition, we should not be surprised that they are both fast and introspectively translucent. The similarities an agent thinks are important—the ones she appeals to, for example, in attempts at moral persuasion or moral education—may well not be the ones that have actually driven her judgment. Moreover, judgments of permission and prohibition are gradient—acts are more or less permissible, or more or less forbidden—in the way we would expect, if response depended on comparison to stereotypes or prototypes, rather than to discrete general principles.

## 7.5   Moral Apprentices

So perhaps reactive moral judgment depends on pattern recognition learned from exemplars, rather than the application of tacit general principles, as both Stephen Stich (1993) and Paul Churchland (1996) suggested some time ago. Furthermore, an exemplar-based view of moral judgment is independently plausible. My earlier chapters will have primed the reader with the general line of argument here. Children are born into a world rich with normatively appraised acts and agents. Other agents are persistent, inveterate moralizers. In pursuing a gossip hypothesis about the origin of language, Robin Dunbar and his group have documented the degree to which much casual conversation is about others and their acts (Dunbar 1996), and much of this is normatively loaded.[6] As soon as language comes online, children are exposed to normative evaluations in stories, from their peers, from their parents' generation. The narrative life of a community—the stock of stories, songs, myths, and tales to which children are exposed—is full of information about what actions are to be admired, and which are to be deplored. Young children's stories include many moral fables: stories of virtue, of right action and motivation rewarded, of vice punished. Their narrative world is richly populated with moral examples. In addition, in some cultures there are systematic attempts to prime and reinforce young children's prosocial emotions. Barry Hewlett describes such practices amongst the !Kung, with !Kung mothers teaching very young infants to share food and beads amongst their sharing network (Hewlett et al. 2011, 1175).

Moreover, children explore, experiment on, and try to manipulate their social life. They try to influence others by moralizing themselves, and they are forced to respond to the attempts of others. Children develop in an environment saturated with local, case-by-case normative evaluations. Thus a multitude of particular experiences, annotated with their moral status, acts as input to a pattern-recognition learning system. Moreover, in contrast to skill transmission and language, children are often explicitly instructed about religious, moral, and legal norms. One famous version of the poverty-of-the-stimulus argument in linguistics turns on the idea that children lack explicit negative information. They get no negative data. Ungrammatical sequences are not produced and labeled as deviant; they are just not produced at all. Thus we need to explain what blocks the acquisition of overgeneralizing grammars. That is not true of moral education; forbidden acts are both produced and described, and their forbidden character is made explicit. That is true of children's lives, and even more in the stories and narratives to which they are exposed. The principles of moral life may not all be made explicit in moral education, but children are exposed to both positive and negative examples.

Thus the general features of human social learning reappear in the development of moral cognition in children. Moral learning, too, is learning by doing, but in a structured and enriched environment. Children do not acquire information about the moral opinions of their community merely by observing what adults do and avoid doing. Adults talk as well as act, both with one another and with children. They respond with emotional signals to what others say and do, and those responses carry normatively important information. Moreover, children experiment with and manipulate their surroundings. Children's social worlds are full of disputed terrain, especially regarding issues of fair division and responsibility. Children collide with the moral views of their peers, and they attempt to impose their own views on those peers. Few children could learn the norms of fair division from the untrammeled actions of their brothers and sisters. But they have a good chance of learning them from overhearing and taking part in discussions of how spoils are to be divided, for example, "You cut, I choose," and similar rules of division. Moreover, games can also play a role in the acquisition of normative cognition, teaching children about taking turns and the like, as well as distinguishing between an authority-dependent rule (as in the rules of a game, which can be changed by negotiation) and authority-independent rules.

In short, our capacity for moral cognition is acquired through an enriched version of the expertise-by-apprenticeship model that I sketched in section 2.3 and developed in the last few chapters. Moral cognition, like many other competences, is acquired through the interaction of individual cognitive adaptations with an informationally organized learning environment. That said, I agree with the nativists that the acquisition of norms is biologically prepared; but the crucial adaptations are perceptual and motivational. To that idea I now turn.

## 7.6    The Biological Preparation of Moral Development

The moral grammarians take moral judgment to depend on general, abstract moral principles, though principles whose application to particular contexts and cultures will depend on local specifications of harm, equity, kinship, and so forth. In part as a result of thinking that moral cognition depends on the application of general, subtle, and abstract principles to particular situations, the moral grammarians think that norm learning is biologically prepared. Abstract principles like the doctrine of double effect are distant from specific moral experiences. Domain-general learning strategies, the thought goes, are unlikely to take a three-year-old from recognizing that her family disapproves of a set of specific acts to the recognition, say, that harms by commission are worse than harms by omission. The moral grammarians conclude that we are adapted for moral thought. Those adaptations accelerate development and make it more reliable. While we genuinely learn about our moral universe, we can learn about that universe more rapidly and reliably because our minds are specifically adapted for the learning task. In turn, those adaptations for moral learning damp down cross-cultural variation in moral systems. I agree that moral development is robust because we are biologically prepared for moral education, but I think that preparation consists in the organization of our developmental environment, through specific perceptual sensitivity, and by our prosocial and commitment emotions.

I have already argued that a child's moral development normally takes place in a prepared environment. But we are also prepared for moral education by the phenomena we find perceptually salient and emotionally arousing. Nativists often model the situation of the first language learner as that of the radical interpreter: a hypothetical anthropologist in an alien community, facing the task of coming to interpret the alien language. In the case of language, this model may well be appropriate, for there may be no interesting distinction between being able to fully describe a language and being able to speak it. So we can think of the child's language-learning task as structurally similar to that of the radical linguist. In the moral domain, the ethnographer's task is also descriptive: namely, to describe the norms of a village or community. But there is a difference between being able to describe moral norms and sharing those norms, that is, making them one's own norms. The child's task is not to discover the set of practices governing a particular community. Rather, her task is to join her community, to share rather than describe those norms. She can do so because moral cognition is embodied (Scott 2010). Her own emotional responses, and her sensitivity to the responses of those that surround her, play a crucial scaffolding role in entering the community in which she lives. Moral judgment and emotional response are intimately connected. Prosocial and commitment emotions evolved before moral cognition; they made possible the cooperation and cultural learning that prepare the evolution of explicit normative thought. The developmental timing of normative thought remains controversial, but Tomasello's studies show that human children behave prosocially, co-operatively, and (to a degree) empathetically before they begin school. In development, prosocial emotions prepare, and perhaps make possible, the development of normative thought.[7] The anthropologist's project is not a helpful model of the child's project.

Thus I am allied with a Humean revival: we are perceptually tuned to the emotions and emotional responses of others, and to our own emotional responses to those others. We are typically aware of, and emotionally responsive to, others' distress. We do not just respond emotionally; we notice our own emotions by some internal analogue of perception. These features of emotion are the launching pad for the Humean revival in psychology and philosophy known as sentimentalism.[8] We respond positively to kindness; we are aware of our own positive response, and we convert that visceral reaction into a normative judgment. Moral norms grow from our dispositions to respond emotionally in characteristic ways to stereotypical stimuli: for example, with disgust to body products and rotting food, or with distress and discomfort to the suffering of others (or at least of those in our group). These emotional responses make certain types of event salient. As Nichols (2005) sees it, putative norms that line up with these characteristic responses are more stable. They are more likely to be endorsed, respected, and taught than arbitrary norms.

Moreover, such norms are much more learnable. Such entrenched, automatic responses increase the salience of situations and actions to children. Normal children, for example, notice when their playmates are distressed. Their own emotional responses to those emotions are motivating. Distress is unsettling to many children. Our emotions make certain types of actions and situations salient: we notice the emotions and reactions of others, and that in itself will focus moral search space; the array of candidate norms that the child considers. In thinking about the content of moral norms, children will tend to think of human emotions, reactions, and the stimuli that engage them. Children (and adults) notice these phenomena and are motivated by them independently of, and before, moral education. So a child's evidential base includes not just patterns in others' actions but also information about others' emotional responses to those actions, and information about that agent's own emotional response (Scott 2010). It is easier to learn a moral norm about the effect of actions on humans than one enjoining respect for the autonomy of cabbages. We notice act-effect-response syndromes: we notice that shouting at people tends to make them upset and to respond aggressively themselves. As a result of the salience of these syndromes to human perception, whether one should shout at others is up for consideration.

In this picture, then, we are biologically prepared to develop moral cognition, not because our minds are prewired to acquire moral concepts and principles, but because moral cognition is a natural development of our existing emotional, intellectual, and social repertoire. The idea is similar to a familiar response to linguistic nativism. Perhaps our minds are not adapted to use language by having language-specific information built into us. Rather, languages are adapted to be used by agents with minds like ours, and their structural similarities follow from that (see, e.g., Deacon 1997; Christiansen and Chater 2008). While intriguing, this idea remains controversial in linguistics. But the parallel suggestion about moral cognition is compelling. Our suite of emotional reactions—especially those concerned with reciprocation, sympathy, empathy, disgust, and esteem—shapes and constrains, though certainly does not determine, moral cognition. According to this view, moral cognition develops from an interaction between emotions, exemplar-guided intuitions, and explicit principles.

Emotions are important in priming moral response at a time and in biasing the set of norms that a culture endorses, sustains, and transmits. Mapping the relationships between gut-feeling morality, explicit moral reasoning, and emotion is a challenge to every model of moral cognition (some options are nicely mapped in Huebner, Dwyer, and Hauser 2009). But it is important to recognize a role for explicit principles. While I sympathize with the sentimentalist family of views, its defenders sometimes underplay the role of explicit moral reasoning. As we have seen, Jon Haidt defends a version of the Humean picture in which intuitive moral judgment is the reflection of our emotional responses of aversion and approval, and in which conscious moral reasoning is often a post hoc response to intuitive judgment (Haidt and Bjorklund 2008). In this picture, conscious moral reasoning is not always confabulation. Especially in social interaction, conscious reasoning can determine moral judgment. But judgment is typically a result of emotional response.

Hauser rightly thinks that this is an overly bottom-up picture of judgment. Moral vegetarians typically come to feel disgust for meat, and that is a cause rather than a consequence of their moral convictions.[9] Emotions sometimes drive moral appraisals, but sometimes appraisals drive emotions. Indeed, moral response always piggybacks on some form of perceptual or cognitive recognition one's situation. Emotional response is not typically dumb and reflexive. So moral vegetarianism and similar examples form an important set of cases. They show that over time, moral appraisal influences moral emotions. Moral appraisal is not just the rationalization of responses we have before and independently of the appraisal. Such examples show that moral cognition (unlike, perhaps, language) has no "critical period" that fixes our systems of moral response by late adolescence and does not significantly change in the life of the individual thereafter. These examples show two-way interaction between reflexive and reflective morality. Over time, to some degree, the reflective morality of an agent rewires his or her reflexive, reactive morality. Moral vegetarianism and similar examples show that we need to identify the ways in which top-down reasoning affects emotions, exemplars, and especially generalization from exemplars. A key form of moral argument is to try to induce others to see similarity relations between cases they already evaluate morally in a distinctive way and the case under dispute. And these arguments sometimes work.

Thus I see that the relationship between the tacit and the explicit is akin to the relationship seen with many skills, and thus it further supports an extension of the apprentice model to this case. A skilled craftsperson has a good deal of explicit information at his or her fingertips: rules of thumb, the lore of the trade. This explicit, articulated, detachable information coexists and interacts with pattern-recognition capacities; well-tuned habits; information that can be made explicit, but only with the right prompts; know-how. Often explicit principles take time to be smoothly integrated with fluent practice; often they can only be partially extracted from that practice. The distinction between explicit and tacit is not sharp: a cabinet-maker may be able to explain, say, the reasons why she rejected one source of raw materials in favor of another, but only slowly and partially, reconstructing the decision rather than reporting on it. Likewise a skilled birder can probably decompose the jizz of a raptor into some explicit components about glide, wing beat, and habitat. But again this is likely to involve some mix of reconstruction and report. Moral competence is like this: an amalgam of explicit principles, rules of thumb, know-how, and prototypical representation.[10] This composite is manifest in moral argument, which certainly involves explicit principles but also often involves trading examples through which each agent attempts to induce the other to see the contentious case as similar to other cases about which the agents already have a well-formed view.

## 7.7   The Expansion of Cultural Learning

In brief, the picture I present here does accept that the development of moral cognition is supported by our evolved biology, but it does not invoke domain-specific principles on the model of grammar. That might be too minimal a view of our biological endowment.[11] I have argued that children develop their moral intuitions about, for example, fairness on the basis of prototypical representations of fair and unfair actions. Moral judgment depends on pattern recognition guided by paradigm instances, so given that a child has the concept of a moral transgression in these paradigm cases, perhaps it is not hard to work out which acts count as transgressions. But perhaps acquiring that core concept is the crucial challenge to a nonnativist view of morality.

Richard Joyce, for one, does argue that this concept of a moral norm poses the key learning problem. In the absence of an explicit account of moral concepts and of what is required to learn them, it would be premature to reject Joyce's minimal but still domain-specific nativism. I am not convinced, but no matter. While some form of minimal nativism remains an option, the point of this chapter is to illustrate the way we can extend the basic model of chapters 1 and 2. One extension of cultural learning is an increase in the fidelity and bandwidth of the transmission of craft skill and ecological expertise. The last two million years have seen an extraordinary expansion of just such bandwidth and fidelity. This expansion depends on (i) changes to hominin life history (resulting both in longer periods as a juvenile and in more experienced and knowledgeable adults); (ii) the expansion of material culture (resulting in a learning environment more richly seeded with examples, partly processed raw materials, toys, and other props); (iii) changes in the social environment, most obviously explicit teaching and other forms of information pooling; and (iv) the evolution of specific adaptations for cultural learning (some of which bring with them information technology such as language and depictive representation). In some cases, we probably have domain-specific information wired in as well, giving cultural learning a head start. The basic principles of folk physics, for example, are probably wired in, thus accelerating the acquisition of many technical competences.

A second expansion occurs in the extension of the scope of cultural learning to new domains. That extension has been the focus of this chapter. I assume that the first users of Oldowan technology had no norms to learn. *Sapiens* children are born into a snake pit of normative and customary expectations, and serious learning failure will doom them. A minimal nativism of the kind that Joyce suggests may aid their learning journey, but the point of this chapter is to illustrate the power, even in this domain, of the hybrid learning strategies that explain expertise transmission. Children learn the norms of their social world by acting in their social world; they learn by doing. But that learning is structured and sustained by cultural tools: fables, moral stories, the myths of their religion; the heroes and villains of their culture; the games they play. They are born into communities that have extensive normative vocabularies, and these help make salient certain categories of act, person, and situation. Perhaps in contrast to skill acquisition, in many cultures, explicit instruction plays a central role. Children play games, and games have rules. Learning the rules of play may scaffold their mastery of the distinction between norms and regularities. They engage in moral play, both in the sense that many childhood games have clear role divisions between heroes and villains and (probably more importantly) in the sense that social play gives children practice in low-stakes moral disagreement and moral negotiation.

These engineered learning worlds have their developmental effects in conjunction with individual cognitive mechanisms of emotional recognition and emotional response. These mechanisms make some aspects of the child's social world salient and mask others. Thus I think that our minds are adapted for moral learning. But here, as elsewhere, evolutionary change proceeds by tinkering. We are adapted for moral learning through tweaks to preexisting adapted learning environments and through tweaks to preexisting capacities to recognize and respond to the emotions of conspecifics. Nothing fancier was needed to make us persistently moralizing and occasionally moral animals.

# 8 Cooperation and Conflict

## 8.1 Group Selection

I began this book by noting the extraordinarily rapid and extensive transformation of the hominin lineage once it had separated from that of the great apes. But hominins contrast with their closest living relatives not just in their phenotypes but in the evolutionary mechanisms that explain those phenotypes. Chimp life involves a significant amount of social learning, and as a consequence, multigenerational behavioral traditions probably affect both social signals and perhaps foraging techniques (see, e.g., Laland and Galef 2009, esp. chaps. 3–5). But these behavioral traditions are not high fidelity, historically deep, or transformative. Most of their learning is individual learning. Chimps are adept at trial-and-error learning, and as they explore their territories, they learn a lot about the resources those territories offer, and the dangers they impose. Some of this learning is socially mediated. Young chimps forage for years with their mothers, and they learn about foraging targets and, perhaps, foraging techniques in part through this association (Whiten 2011). But social learning, including hybrid learning, plays a limited role in their lives. Though they learn through social interaction, their minds have not been transformed to share information. Thus one telling difference between the chimp species and humans is that from a very young age, children point demonstratively, to indicate to their audience an item of mutual interest. Chimps point acquisitively, to indicate something they want. But they do not point demonstratively, and they find it difficult to learn to exploit human demonstrative pointing (Tomasello 2008, 2009).

The world into which chimps are born—the locale in which they grow up—is inherited from their parents' generation. Chimps tend to live in reasonably stable, well-defined territories. They are locally mobile as they forage through their range, but they do not move broadly across landscapes. But chimps do not transform the territories in which they live. They forage, and build nests, but they do not reshape landscapes, as beavers do, or organize and filter their interactions with their world, as termites do. As the generations go by, the chimps' world changes, but it does so through external impacts on their world, not because chimps themselves change their world in ways that tend to accumulate over time. They live in a world as they find it, rather than a world as they make it.

Moreover, chimps act for themselves alone. In most circumstances, we can understand their behavior by modeling them as maximizing their individual expected fitness.[1] We can observe some cooperation in their lives: males form coalitions against males from other groups and in pursuit of dominance in their own group.[2] Males hunt together (though Tomasello [2008] argues that this is an every-chimp-for-himself scramble, rather than a cooperative activity). Michael Tomasello and Felix Warneken show that chimps have some prosocial willingness to help others in small ways (Warneken and Tomasello 2006, 2009; Tomasello 2009). But this collapses in the face of temptation, and collaborative activities (especially those of male coalitions) seem to be instrumental cooperation in pursuit of self-interest. Warneken shows that while young children find collaborative activities intrinsically rewarding, young chimps do not. Chimps are willing to engage in joint action, but only for instrumental ends (Warneken, forthcoming). Overall, and perhaps with the exception of a taste for revenge (Jensen, Call, and Tomasello 2007b), chimp lives can be quite well modeled as *Pan economicus*. They are rugged individualists.

In the argument of the book so far, I have underscored two profound differences between hominin and pan evolution: social learning and niche construction. Hominins organize their own environment and that of their succeeding generation. I have suggested that skill transmission began as a by-product of adult activities: the learning environment of the earliest young hominins may have been adaptive, but it was not adapted. As hominin social life became richer, with the young dependent for longer periods, and as the informational load on hominin lives increased, information flow began to depend on both specific cognitive adaptations and an adapted learning niche. No doubt the specifics of that learning niche have varied over space and time. But downstream niche construction is a core mechanism of hominin evolution.

Moreover, change tends to accumulate over time. That is certainly a dominating aspect of life in the Holocene, as the material and biological environment became increasingly human made. Farming life, and ultimately urban life, replaced foraging. But the roots of accumulating change go deep into the Pleistocene. For one thing, as foragers, humans reshaped many of their biological environments. They altered animal community structure by displacing and persecuting rival predators and by reducing the populations of favored prey animals. They changed vegetation patterns, and hence the distribution and abundance of animal resources, through their effects on herbivores, as well as more directly, for example, by fire-stick farming (Bliege Bird et al. 2008). For another, the world into which children were born became richer and richer in material and symbolic technology. Language, to take the most obvious example, is an immensely rich system of classification; the distinctions marked by vocabulary and idiom reflect a culture's accumulated experience of its environment and its most salient features. As I have noted before, dating the evolution of language is still guesswork. But a child born, say, 100,000 years ago into a linguistically rich forager's world experienced a very different environment from a child born, say, 1 million years ago into a world without such a language. That difference resulted from cumulative niche construction, not an externally driven change in the world.

Those who have developed the theory of niche construction and shown its importance to evolutionary change have seen it as a form of inheritance. Organisms that alter their own environment often change their descendants' environment, too. Arguably, these effects are especially salient in hominin evolution. Children inherit a world and a set of material and cognitive tools from their parents, not just their genes. As a consequence of the interactions between their parents and their parents' world, children inherit an altered set of selective forces and developmental resources (Laland et al. 2000; Odling-Smee et al. 2003; Laland 2007). However, if this is inheritance, it is inheritance with a difference. The children of a particular foraging band inherit genes only from their parents, and hence there will be differences within the band that depend on differences in genetic inheritance. Different individuals have heritable differences in phenotype and fitness. Is the same true of their inherited niche? The children in a particular foraging band will not be born into strictly identical learning environments, with the same set of cognitive tools, material resources, and engineered relations with their natural world. But it seems quite likely that many of the differences will be noise, perhaps all of them.[3] It is not obvious that this form of inheritance results in heritable differences in individual, within-group phenotype and fitness, differences of the kind visible to selection (Dawkins 2004). However, they might result in stable phenotypic differences *between* different foraging bands, perhaps adapting one group, but not its neighbor, to pursuit hunting (to revert to the example discussed in sec. 3.5). So if niche construction is an inheritance mechanism generating selectable variation in hominin evolution, the population on which selection acts seems likely to be a population of bands rather than a population of individuals.

A similar suggestion emerges from a consideration of cultural learning. As many evolutionary theorists have pointed out, evolution by natural selection depends on the existence of parent–offspring phenotypic similarity, but not on the specific mechanism through which those similarities are generated (Mameli 2004; Godfrey-Smith 2009). So intergenerational cultural learning is often seen as an inheritance system. Agents that regularly learn from their parents will resemble their parents more than they resemble other members of the population. If those differences are relevant to fitness, selection will act, and the lineage will evolve (Boyd and Richerson 1996; Avital and Jablonka 2000; Laland and Brown 2002; Shennan 2002; Jablonka and Lamb 2005). But thinking of cultural learning in this way does presuppose that children learn distinctively *from their parents*, not from parents together with other members of their parents' generation. If the model I suggested in section 2.3 is on the money, skill transmission in hominins probably began this way, as young hominins accompanied their mothers and explored a world whose characteristics were largely determined by their mothers' foraging routines and skills.

In contemporary and near-contemporary worlds, some cultural information flows vertically. As noted in section 2.3, in many cultures, the skills of weaving and other handicrafts flow from mother to daughter. There is also some suggestion that herbal lore (used in folk medicine) flows in families rather than through a group as a whole (Hewlett et al. 2011, 1174). But much does not. Information flow became part of a more cooperative world. Reproductive cooperation increases the range of models—information sources—to which children are exposed (Burkart et al. 2009). So too does the tolerance of youthful satellites in multi-adult cooperative activities (Marlowe 2007). Moreover, children forage together in many forager cultures and, in doing so, have the opportunity to learn from their peers and also, indirectly, from their peers' parents. This is especially true after children grow from toddlers towards early adolescence (Csibra and Gergely 2011; Hewlett et al. 2011). For example, on a fishing trip, a cohort of children will carry nets, lines, and baskets that their parents have supplied (Bock 2005). The more cooperative and interactive the local group, the more information flows through many-sender, many-receiver networks. Each adult becomes a source for many children; each child has access to many sources. At the limit of this process, within a local group, information is pooled at a generation and flows diffusely, through many overlapping routes, to each child in the next generation. Information pooling thus erodes the potential for cultural learning to form an individual-to-individual inheritance channel acting in parallel to genetic inheritance, with similar evolutionary dynamics. It makes cultural learning less likely to result in stable phenotypic differences within a group, differences that are potentially relevant to fitness and thus visible to individual-level selection.

However, as David Wilson, Elliot Sober, Peter Richerson, and Robert Boyd have pointed out, the same information-pooling mechanism makes the bands themselves potential inputs to selective filtering. The metapopulation of bands is itself a Darwinian population (Soltis et al. 1995; Sober and Wilson 1998; Richerson and Boyd 1999). Selection on bands is potentially powerful if bands differ from one another in ways that are relevant to the fitness of those bands, and if those differences are transmitted to groups that form as bands fissure.[4] Group selection is made more powerful by any process that leads to groups being internally homogeneous, and different from other groups. The combination of information pooling and cross-generational learning has just these effects. So perhaps the expansion of various forms of cooperation within local groups led to the formation of a metapopulation of local hominin groups subject to natural selection. As a consequence of cumulative niche construction and cultural learning, human groups differed, one from another, in ways relevant to those groups' fitness. Selection shapes human phenotypes by filtering individual humans, rewarding some phenotypes and dooming others. But it shapes individual human phenotypes, and the characteristics of the groups within which humans live, by filtering human groups, rewarding some of those groups, and dooming others to extinction.

Researchers disagree about the right ways of making this idea precise and applying it to real data. As Samir Okasha, in particular, has made clear (building on Damuth and Heisler 1988), we can measure group fitness in two ways. One group can be fitter than another if it pumps more individual hominins into the next generation (multilevel selection 1). Or it can be fitter because it is more apt to produce descendant groups (multilevel selection 2) (Okasha 2006; Godfrey-Smith 2009). Many of the formal models of group selection have measured fitness in the first way, and that includes formal models intended to support the picture of cooperation and its evolution considered in the next section (Choi and Bowles 2007; Gintis 2008). As we shall see, however, much of the anthropological support is most naturally interpreted as thinking of group fitness as group survival.

In section 5.3, I introduced the idea that human cooperation in large groups depends on the fact that many humans have the psychology of strong reciprocation. I now return to this idea. Its chief defenders, Sam Bowles and Herbert Gintis, argue that this psychology evolved through group selection and in an environment of intense, often violent competition between groups. While I think that group selection has likely played an important role in hominin evolution, I do not think we need group selection to explain the evolution of the capacities that sustain human cooperation. In the next two sections, I take up this model and, in the final section, use it to explore the transition from the Pleistocene to Holocene social life.

In the last decade, Peter Richerson and Robert Boyd have increasingly emphasized the strangeness of the Pleistocene world. It is, as they note, "a world queerer than we had supposed." But difference is symmetrical. The more the Holocene contrasts with the Pleistocene, the more we face a challenge in explaining how groups, minds, technologies, and mores that were adapted to the Pleistocene coped with the Holocene. I argue that we should be surprised that human mechanisms of decision, motivation, and cooperation were robust enough to maintain functionality in the Holocene. The Holocene is an underappreciated explanatory challenge.

## 8.2 Strong Reciprocity and Human Cooperation

Sam Bowles, Herbert Gintis, and their collaborators make three linked claims about human cooperation and its evolution.[5] First, they develop a thesis about human psychology. Humans are (typically) strong reciprocators, or, as they sometimes say, humans have social preferences. They are sensitive not just to the personal economic consequences of their decisions but also to the effects of those decisions on others. Thus they enter interactions disposed to cooperate if they expect others to, perhaps even at some economic cost. But they are also "moralistic." Humans are often angered by free riding, sometimes even when these failures have no effect on their own returns, and are often willing to pay to punish those who fail to cooperate. As Bowles and his allies see it, an array of results from experimental economics show that neither cooperation nor punishment is merely a prudent investment in the future behavior of partners.[6]

Second, they argue that the psychology of strong reciprocation could not have evolved by individual selection. Repeated interaction favors cooperation. But even when agents do repeatedly interact, if they do so in groups of moderate size, it would not pay to be a strong reciprocator. Cooperation is not stable in larger groups—free riders invade—unless free riders and only free riders can be targeted for punishment. Thus in four-player public-goods games without punishment, cooperation declines as free riding invades (see, e.g., Gächter et al. 2010). Dispositions to identify and punish free riders will not evolve by individual selection, for it is expensive both to identify free riders and to punish them. The only credible mechanism for suppressing first-order free riding creates a second-order free-riding opportunity, one that would itself result in the decay of cooperation in large groups. In the view of Bowles and his colleagues, strong reciprocation is a form of altruism, not just in the psychological sense but also in the evolutionary sense. The fitness benefits that explain the trait do not accrue solely to the agent that has the trait.[7]

Third, Bowles and Gintis argue that the psychology of strong reciprocation both could and did evolve by selection on groups. They have developed a family of formal models that they take to show the theoretical plausibility of a group-level mechanism. The models show altruistic cooperation evolving (i) when we make plausible estimates of the degree of within-group inbreeding, and hence the genetic differentiation we might expect between groups; and (ii) when we make plausible estimates about the advantage that altruism confers on one group in its conflicts with others. They combine their model-based argument with a historical scenario of frequent, often lethal, interband hostility in the Pleistocene.

The overall picture, then, is that the psychological profile that makes behaviorally modern cooperation possible has evolved relatively recently, via group selection in a Pleistocene world of intergroup tension that often escalates into raiding and violence. This psychological profile incorporates some of the most recently evolved and sophisticated features of the human mind, in particular the cultural transmission and internalization, by means of full language, of norms. In contrast, I suggest that the bedrock psychology of cooperation is more ancient, forming part of the environment in which language, normative thought, conventions, and institutions evolved. The capacities that make cooperation possible evolved because they paid at the individual level, not because agents lived in a warlike world. Thus I argue for an inversion of Bowles's and Gintis's picture. Cooperation evolved earlier, by individual selection and in an environment of relatively peaceful intercommunity relations. But conflict did come. So how did these prosocial dispositions survive, and continue to shape prosocial behavior, as the social world became more systematically and lethally violent, more vertically complex and differentiated, and much less egalitarian? These changes all seem to increase the attractions of defection. We ought to expect a late Pleistocene or early Holocene cooperation crisis to have developed. The survival of cooperation through the transition to the larger-scale, less-egalitarian social worlds of the Holocene needs explanation.

So, three claims are the target of my discussion: (i) humans are strong reciprocators; (ii) strong reciprocation could not evolve by individual selection, because of the costs of monitoring and punishment; and (iii) strong reciprocation could and probably did evolve because of intense competition for resources between groups in the Pleistocene. I take up the first two in this section, accepting the first with some reservations, but rejecting the second. I reserve the third claim for the next section.

Bowles and Gintis are probably right to read behavioral economics as showing that most people are default cooperators. It is hard to explain such phenomena as the voluntary provision of public goods on other assumptions, and the evidence from experimental economics is even more persuasive. Even so, Ken Binmore is not convinced: he tries to explain away the experimental-economics evidence by arguing that agents need to learn about the payoffs of the games they are in. Agents are self-interested, but they begin by using cognitive and decision-making heuristics that are learned in the context of repeated and public interactions. Agents learn that these do not pay the special and unnatural case of anonymous interactions, and so we see, for example, the erosion of cooperation in public-goods games without punishment (Binmore 2006). Gintis is convincing in reply, arguing that empirical evidence from experimental protocols shows that subjects understand the game, and that this interpretation is supported by their rational response to alterations in the payoff structure. For example, as punishment becomes ineffective or too expensive, agents stop punishing.[8] Finally, in contemporary, large-scale societies, we have many one-off, effectively anonymous interactions. So the idea that in the early stages of these interactions agents do not understand their circumstances is implausible (Gintis 2006).

Even so, we need to be cautious about the claim that humans are typically strong reciprocators. Humans exhibit individual variation and phenotypic plasticity. Perhaps even more importantly, Simon Gächter and his colleagues have shown important cultural variation in agent decision making, especially with respect to punishment and response to punishment (Gächter and Herrmann 2009; Gächter et al. 2010). Thus strong-reciprocation psychology probably comprises a complex of capacities that were assembled incrementally and can still partially disassociate from one another. Cooperative defaults are sensitive to the form of cooperation; young chimps have cooperative defaults with respect to manipulations of the physical environment, but not with food or information sharing (Tomasello 2009). Likewise, cooperative defaults can be more or less fragile in the face of temptation. Punishment, too, is not monolithic. As I noted earlier, Gächter and his colleagues show that in some agents, willingness to punish disassociates from willingness to cooperate. Some uncooperative agents punish cooperators (this is known as "antisocial punishment"). And almost certainly, retaliatory dispositions began as a response to personal injury, before, in some agents, generalizing to a willingness to retaliate on behalf of others.

Despite these reservations, humans do indeed seem to have robust dispositions to cooperate and to retaliate. But these psychological dispositions might have evolved through the individual advantages they confer. For many hundreds of thousands of years, our ancestors were long-lived, with good memories, and were parts of modestly sized and relatively well-defined groups. Moreover, as I have argued at length, life in those groups afforded many opportunities for profitable cooperation. In the hominin

lineage, the ecological, social, and psychological preconditions for cooperation based on reciprocation have long been met. Given this picture of
hominin history and ecology, surely it paid to be a default cooperator and
to vengefully punish free riding. Agents with those dispositions secured
the advantages of direct and indirect reciprocation. Punishing free riding
is not free, but it advertises your own commitment to cooperation, both to
onlookers who see you act and to a wider circle who hear of it.

Bowles and Gintis are highly skeptical about reciprocation-based,
individual-advantage models of the evolution of strong-reciprocation psychology. They are skeptical in part because they think reciprocation is likely
to fail when it is most important. The shadow of the future is least certain
just when it is most necessary, namely, when a group faces crisis:

> The probability of one's contributions being repaid in the future, however, decreases
> sharply when the group is threatened, since the probability that the group will dis-
> solve increases and hence the incentive to cooperate will dissolve. Thus, precisely
> when a group is most in need of prosocial behaviour, cooperation based on reciprocal
> altruism will collapse. Such critical periods were common in the evolutionary history
> of our species. (Bowles and Gintis 2003, 434)

I argue in the next section that Bowles and Gintis exaggerate the role of
intergroup conflict in the Pleistocene. But even if that is a mistake, Pleistocene bands could not have routinely dissolved in the face of crisis: of war,
famine, or pestilence. That would be true only if agents could improve their
chances of survival by going it alone, and that they could not do. Especially
if the environment really was one of hostile interactions between groups,
the members of a group must cling together. Males, especially, will have
few prospects of migrating into another group, and an isolated individual
or a small nuclear group would be easy pickings, both to other humans, if
they are as hostile as this picture supposes, and to predators. Life is tough
indeed for low-ranked individuals in primate societies (for a vivid picture,
see Sapolsky 2002). Yet they do not try to go it alone, presumably because
it would be suicidal to do so.

Perhaps most critically, Bowles and Gintis think the problem of scale
is intractable for individual-selection accounts of cooperation. As noted
earlier, cooperation is unstable even in modestly sized groups, unless free
riders can be specifically targeted. That presupposes that such free riders
can be identified and, once identified, punished. I do not see identification as a serious problem: I have already argued that in small, repeatedly

interacting band-sized groups, agents will become well informed about one another just by direct observation and memory. This effect is magnified by gossip, once such communicative capacities evolve. Of course, gossip is not perfectly honest and reliable, but as I argued in chapter 6, multisender, multireceiver networks are well insulated against deception, and participating in those networks brings benefits of reciprocation and information pooling. No social environment is perfectly transparent, but in the social environments in which strong-reciprocity psychologies evolved, agents were awash with information about their peers.

It is true that sharing information via gossip is indeed an $N$-player cooperation problem. But the costs and benefits of informational cooperation do not mirror those of ecological cooperation. Costs are lower; benefits are higher. The benefits of information sharing increase as $N$ increases, for Condorcet-like effects increase the reliability of consensus. And the more that gossip is multisourced and multitargeted, the less likely it is to be deceptive manipulation. Moreover, in contrast to sharing material resources, the costs of sharing information do not increase with the number of agents aided. But the number of reciprocation sources does rise with informational targets. You can help more people—and hence legitimately expect help back from more people—for a fixed cost. Finally, Nowak and Sigmund point out that experimental evidence shows that agents are sensitive to others' reputation and expect others to be sensitive to their own reputation. Thus these theoretical considerations have experimental support. Public-goods games and their relatives show indirect reciprocation in action. Reputation attracts cooperation: generosity in helping can pay. Moreover, agents understand that fact and respond by investing in reputation—if other agents have information about past behavior, agents behave more cooperatively (Nowak and Sigmund 2005).[9]

So the problem of identifying free riders is a pseudo-problem. In this respect, the formal models of the role of reputation in the evolution of cooperation are seriously misleading. For example, in image-scoring models, an agent who has a good reputation is more likely to be treated cooperatively by others, and agents' reputations depend on others' information about their past interactions. This can be made to generate a problem. How can an onlooker tell whether I am defecting against a third party or punishing a third party? To score my image, it can seem as if that onlooker would need an implausibly accurate and complete history of my previous interactions, and of other parties to my interactions. The modeling literature develops different versions of this problem (Nowak and Sigmund 2005; Bowles and Gintis 2006). But here the model fails to capture its target. As the models represent these interactions, onlookers see no intrinsic difference between one agent defecting against a second, and that agent punishing the second. In real social environments, these are very different. For example, communication among those interacting will be very different. The defector will be attempting to persuade his victim that he is in fact cooperating; in the case of punishment, the agent will be denouncing his target. Denunciation will not tell us whether the punishment is just. But onlookers will nonetheless see two very different interactions.

But while the identification problem is a mirage, Bowles, Gintis, and their colleagues are right in thinking that punishment has costs.[10] Chapter 5 was largely devoted to exploring ways in which such costs bring individual benefits, by securing entry into alliances and by securing trust once such alliances are formed. Fitness benefits flow to individuals through successful cooperation, and securing these benefits often depends on being chosen by others. As first Robert Frank and more recently Paul Seabright have argued, often the best way of seeming to be a good partner in cooperative enterprises is to actually be a good partner in such enterprises (Frank 1988; Seabright 2006, 2010). A default cooperator and willing retaliator is an attractive companion in many projects. Such a partner will not rip you off and will help you defend against third-party cheating and aggression. Given the benefits of cooperation and the importance of partner choice, the costs of punishment are worth paying.

Moreover, if the costs of punishment are too high for individuals to improve their fitness by paying them, appealing to selection at the level of the group is unlikely to help. Gächter and Herrmann point out that material punishment—any punishment that requires resources to be used—imposes a group-level tax. Punishment imposes a cost on those punishing and on those being punished. Thus the total resource envelope created by collective action is reduced by the costs of punishment. Indeed, in some of Gächter and Herrmann's experimental games, collective action sustained by punishment was no more profitable than collective action eroded by widespread free riding. Thus this tax can eliminate the group-level benefit of higher levels of cooperation bought about by punishment (Gächter and Herrmann 2009). Of course, this example might be an artifact of the experimental setup. But Gächter and his colleagues have also shown that antisocial punishment is widespread in some cultures. Those who cooperate less pay to punish those who cooperate more, probably because they blame cooperators for punishment that comes their way (Gächter and Herrmann 2009; Gächter et al. 2010).

This work suggests that punishment may stabilize cooperation only if it is seen as punishment, and only if it is seen as fair. The potential costs of punishment include resentment, strife, and retaliatory punishment. We should remember that the psychology of strong reciprocation did not evolve in an initial population of economists—of coolheaded, rational maximizers. Rather, the baseline population probably consisted of agents who had some prosocial tendencies to cooperate (as chimps do) but who were also impulsive, finding it difficult to resist acts with immediate payoffs. Moreover, the psychology of retaliation was probably already well established in this baseline psychology. The dispositions that make one's threats trustworthy can evolve early; they pay even before the evolution of a richly cooperative world. Antisocial punishment is naturally interpreted as an expression of this phylogenetically deep and developmentally well-entrenched motivation to retaliate against injury. Antisocial punishment shows that punishment might well have produced retaliation, rather than prudential conformity to cooperative norms (as some of the models of the evolution of cooperation suppose (Boyd and Richerson 1992; Boyd et al. 2005).

I do not doubt that punishment was important. For humanlike levels of cooperation to evolve, our ancestors must have had ways of sanctioning those who did not cooperate. But those sanctions cannot have been expensive, or if they were, they cannot have been needed frequently. Otherwise cooperation would not have paid, and that is true whether selection acted on individuals, groups, or both. In my view, then, default cooperation and the desire to retaliate probably evolved by individual selection, perhaps quite early in the hominin trajectory, at least by *Homo erectus*, quite likely much earlier. Multilevel selection probably played an important role in human evolution. But cooperative psychology evolved at least in part because cooperation paid individuals in the groups in which we lived, rather than wholly or largely by group selection. Accordingly, I now turn to the Bowles-Gintis picture of our Pleistocene past and the lessons they draw about altruism and group selection.

## 8.3   Children of Strife?

Chimps live in quite stable territories, and while females can migrate between groups, males treat one another with hostility; indeed, opportunistic but lethal violence flares up when a lone male of one group is seen and targeted by a multimale party from a neighboring group. A three-to-one advantage suffices for a coalition of adult males to have a good chance of killing a single male with almost no risk to themselves. Richard Wrangham (1999) builds an imbalance-of-power hypothesis on the basis of chimp natural history to explain the origins of war in the hominin lineage. War began with the coalition-based violence of one group toward a neighbor that is down on its luck and short of adult males. An imbalance of power creates the opportunity to push a weak group out of its territory and seize its resources, perhaps including its females. Neighbors are a threat, so it is always to a group's advantage to pick off neighboring males when it can safely do so. If your group happens to be the smaller, this helps redress the balance; if your group happens to be the larger, the imbalance is exacerbated, perhaps to the point at which takeover becomes a real option.

Bowles and his colleagues clearly find this picture compelling (Choi and Bowles 2007; Bowles 2009). They develop formal models of the evolution of patriotic altruism based on an image of Pleistocene social life as a struggle for limited resources between bands, supporting this scenario with archaeological evidence that suggests extraordinarily high rates of death from intercommunity violence. Our prosocial dispositions were forged in the crucible of war. In Pleistocene life, intercommunity relations were typically tense and difficult, often deteriorating into raiding and war—the Yanamono with their endless raiding emerge as the Pleistocene default. I have no problem with the formal models that show the potential power of intergroup competition. But while the models show that group selection *could* drive evolutionary change, I do not find the archaeological evidence or the basic conceptual underpinning of this picture of the Pleistocene plausible.

First archaeology. Critically, Bowles and his colleagues refer to late Pleistocene and Holocene sites to show that the Pleistocene was a world of violence. But the Holocene does not make a good model of the Pleistocene. Populations were more sedentary; they were larger, more hierarchically structured, and under greater resource pressure. Further, as Bowles and his colleagues note, we cannot easily distinguish the results of between-community violence from within-community violence. This is no minor detail. As Paul Seabright (2010) has noted, cultures that lack top-down mechanisms of command and control are prone to extremely high homicide rates. Finally, as we saw in the discussion of the demography of the Grandmother Hypothesis, skeletal assemblages overcount death by violence (as a proportion of total deaths) because the less-robust bones of the very young and old are less likely to be found.

The crucial point, though, is that even if the archaeological record of very late Pleistocene and early Holocene war and raiding is persuasive (as it probably is), we can find no archaeological smoking gun showing that intercommunal violence was also prevalent in the Pleistocene. Indeed, to the contrary, Dale Guthrie (2005) argues that much Pleistocene art is the work of adolescent males, and in many ways it does reflect the testosterone-fueled obsessions of such artists. But in contrast with more recent art, we have almost no representation of human conflict from the Pleistocene. As Richerson and Boyd point out, this gap seems significant (Richerson, forth-coming). If the Pleistocene world was one in which raiding was rife, surely we would expect young men to project their fantasies of success and ac-clamation onto cave walls, along with the usual vulvas and penises, which abound.

Most importantly, chimp raiding is not a good model for Pleistocene life. First, as Raymond Kelly (2005) emphasizes, the chimp model gets the cost-benefit picture wrong. In contrast to chimps, humans are armed, and interactions with weapons always impose serious risks. For at least the last four hundred thousand years, humans have made and used spears. Even if a three-on-one attack by chimp males on a lone chimp is risk free for the co-alition, it would not be risk free for human attackers. There would be a real risk that at least one would be badly hurt or killed. Moreover, while chimps do not track, ambush, or stalk, foragers do. So an incursion—a raiding pa-trol—into hostile territory risks serious trouble. Indeed, Kelly argues that the danger from ambush is so great that first detection trumps weight of numbers in assessing relative risk in forager raiding. Here he exaggerates. He uses the Andaman Islands as a model, and they were armed with bows that could be fired from concealed positions. But bows were probably available only in the late Pleistocene. Throwing a spear from a concealed position is much harder; it is a full-body act. As Kelly exaggerates the cost-benefit ra-tios in the opposite direction, intercommunity relations in the Pleistocene were probably not as peaceful as he supposes. Even so, the risks of raiding were high, especially when we take into account the importance of local knowledge of terrain, and the fact that human hunters can call on aid from their allies. Allies are likely to be close and to respond quickly, vigorously, and dangerously. Even for larger groups, a decision to raid involves real risk.

Moreover, good relations with neighbors pay important peace dividends. Kelly points out that male chimps, very prudently, tend to avoid the border zones between their own territory and adjacent ones. That imposes a real tax on hostility: territory is exploited less efficiently. Perhaps most importantly, as Ambrose, Boyd, and Richerson all point out, good relations with neighbors is a risk reduction strategy, allowing access to resources and support in the face of local catastrophe (Ambrose 2010; Richerson, forthcoming). Ambrose details !Kung cultural practices that build mutual commitment and support through visits, gifts, intermarriage, intergroup adoption, and the like. Kelly details similar practices between neighboring groups in the Andaman Islands. These cases show that alliance building for mutual support is not free or automatic; rather, it requires consistent investment. But for foragers in marginal and fluctuating habitats, it is likely to be an important survival mechanism. Neighbors are potentially resources, not just threats and targets.

Given a more realistic assessment of costs and benefits, it is far from obvious that even a resource-stressed Pleistocene forager band should take a predatory attitude toward its neighbors. And how often were the resources of such bands stressed? Pleistocene ecosystems seem often to have been out of equilibrium, as Richerson and Boyd note (Richerson et al. 2001; Richerson, forthcoming). Many of these systems were subject to frequent, high-amplitude climatic variation. They seem to have been dominated by disturbance rather than being structured by local competitions in which locally superior forms squeezed out others (for a classic study, see Coope 1994). Perhaps human demography was dominated by disturbance, too. We do not see much evidence of resource intensification in the human ecological footprint until the late Pleistocene, after the transition to behaviorally modern culture. After that transition, as discussed in chapter 3, we see the broadening of the forager resource base in the Upper Paleolithic. It may be that for much of the Pleistocene, hominin home ranges were dominated by disturbance. If so, local groups would have experienced frequent extinctions and population crashes. But they would also have experienced many periods in which the ratio of population size to available resources was quite favorable. On this picture, much human mortality in the Pleistocene would result from sudden, unpredictable, and extreme local events. Our Pleistocene ancestors often died in savage blizzards, brutal storms or floods, exceptionally long and harsh winters, and perhaps even wildfires when rains failed. They did not typically die from a long, slow grind toward starvation and desperation, over months and years, as the available food slowly shrank as the number of mouths grew. Such a grind might indeed result in desperate competition over dwindling resources. Unlike those of early farmers, forager skeletons do not show signs of malnutrition and resource stress (Cohen 2009).

The Wrangham-Bowles-Gintis picture supposes that between, say, 80,000 and 12,000 years ago, as the cooperative mind was fine-tuned, most human groups were at or near carrying capacity (and that is why we have no archaeological signal of steady population growth). That may not be right. Since Darwin, evolutionary biologists have been fond of calculations showing that we would all be up to our necks in elephant dung in short order, unless *something* constrained the growth of elephant numbers. Clearly, resource availability sets an upper bound on those numbers. But it does not follow from these general considerations that in practice, population size is typically close to, and controlled by, that carrying capacity. There has long been a debate in community ecology about the extent to which population size is controlled by local community interactions, by so-called density-dependent factors, that is, factors that press harder on a population as its size increases (Ricklefs and Schulter 1993; Cooper 2003; Sterelny 2006). No consensus has emerged. In particular, though, we certainly have no clear consensus that for the most part, population size *is* controlled by such density-dependent factors. Pleistocene humans might have been like many other animals, with their numbers controlled by external disturbance rather than local competition. After all, we do know that they lived in a world subject to frequent and intense disturbance. To return to the themes of section 8.1, if this is right, selection on groups might well have been important. But the traits favored would have been group-level traits, probably transmitted by cultural learning and niche construction, that supported risk management. So, for example, there would have been selection in favor of cultural practices that supported the collective construction of shelters, hearths, and communal stocks of firewood; practices that disseminated crucial information widely throughout the band, rather than confining crucial skills to a few specialists; or practices that discouraged dependence on a few crucial resources. Traits that helped bands avoid conflicts with neighbors might well have been favored, too.

Finally, what of movement? Many forager bands probably did live in fairly stable local areas. They were mobile, but mobile through a fairly fixed territory. Others had a different lifeway: they specialized on wide-ranging large game, and such foragers had to be hypermobile. They had to follow the game. As we saw in section 3.5, that pattern would have been true of the pursuit hunters of the mammoth steppes (pursuing highly mobile reindeer herds, for example). Many large African herbivores are seasonally migratory, so there probably would have been African equivalents, pursuing the herbivores of the African plains. Such foragers do not have fixed territories to defend or permanent neighbors. Keeping track of the movements of other bands would be extremely difficult, as each would have to respond immediately to the movement of their particular targets. Moreover, peaceful coexistence would be extraordinarily important for them, as herds fused and fragmented in the face of seasonal cycles and annual changes. Life as a pursuit hunter would be precarious enough without always having to take precautions against being caught at a disadvantage by other bands. The chimp-based conceptual model does not fit foragers who based their lives around the pursuit of mobile herds, and this was probably a significant mode of foraging life.

In summary, I have no doubt that relations between groups were variable and contingent throughout the Pleistocene. Given human capacities for organization, resentment, and violence, one way in which human groups interacted with their neighbors was through raiding and war. That possibility may well have been increasingly salient in the late Pleistocene and Holocene as population sizes grew and foragers and farmers, with their divergent interests, were forced to interact. But I doubt that it was anything like the modal mode of interaction through the Middle Stone Age or in the slow and patchy transition to behaviorally modern cultures. Cooperation and altruism are the fuel of war, but not warfare's child.

## 8.4   The Holocene: A World Queerer Than We Realized?

I am skeptical about the idea that the Pleistocene world was dominated by intergroup violence. But the picture is plausible when reframed as a model of Holocene farming and the interactions between farmers and foragers. Paul Seabright (2008, 2010) sees the Holocene in these terms. In his view, the invention of farming forces farming cultures to invest in defensive technology and resources, as they now have fixed, concentrated, partially portable, and sometimes accumulating assets to defend. They become a stable and juicy target to marauding bands, so they begin to invest in village walls, perhaps with a few semiprofessional guards and purpose-built weapons. These investments in defense make such fortified villages dangerous neighbors, both to other farming groups and to any surviving forager cultures in the local area. A group armed against raiders has the means and the temptation to become raiders themselves. So depending on the exact patterns of response, we will see a mixture of expansion and emulation as neighbors become farmers, so that they too have strongholds and the capacity to retaliate.

One of the puzzles about the origins of farming is its rapid spread, and Seabright offers a solution to that problem. A mixed metapopulation of farmers and foragers is not stable. Once the environmental conditions stabilized in ways that made farming possible, farming was bound to establish somewhere (Richerson, Boyd, and Bettinger 2001; Richerson, forthcoming). Once it establishes, farmers swallow their neighbors or force those neighbors to become farmers themselves. Thus farming ripples out from its points of origin into all those regions in which it is environmentally feasible. I agree that the farming-foraging mix is unstable, though I doubt that the mechanism through which farming spreads is mainly a military imbalance that favors farmers. Foragers do not have much to steal (except, perhaps, their own bodies) and hence are not a tempting target. Moreover, in raiding and low-intensity conflict, foraging groups would have many advantages. They are mobile and dispersed. Their demographic core is not a stationary target. Moreover, individual foragers have natural advantages in field craft and use of weapons, for these skills are an automatic by-product of forager life. Farmers, in contrast, would often be vulnerable to raiders and ambushing, as the need to work their fields forces them into dispersed vulnerability in predictable locations. Farmers are tied to their fields, especially in harvest seasons. Once the harvest is in, their food stores would be defensible (though an immensely tempting target). But while still growing, fields of grain are vulnerable, as are the farmers who need to work and protect crops against many nonhuman threats.

However, farmers are bad news for foraging neighbors, for farmers degrade boundary habitat by constant intrusion and use, as well as making their own fields and livestock off-limits. So farming probably spread in part by degrading the habitat of foraging neighbors, nibbling away at the edges of their ranges. Game and plant resources in the edge habitat would be taken by farmers. And that pressure would be relentless, as the farmers would not move on as returns fell. So edge habitat would tend to lose value and support foraging populations less well, inducing some tendency for foragers to withdraw and for farmers to advance. Resource depletion in buffer zones is bad news for farmers, making them more dependent on their farmed resources. But it is worse news for any foraging neighbors they might still have.

This boundary effect will be especially strong if farming populations are larger than those of the foragers whom farmers replace and adjoin. That is quite likely, both because farming taps into a lower level of the trophic pyramid and because, as Weisdorf (2005) argues, farming rates of return increase in the early period of farming. Early on, good-quality land is not in short supply. Moreover, it is easy to find major improvements early in a technological tradition (Kauffman 1995). By the time the transition to farming is taking place, the easily found, large-effect breakthroughs in foraging—distance weapons, dogs, cooperation, seasonal exploitation—have already been found and exploited. Without a really major advance in technological capability, foragers can only fine-tune their existing capabilities, making only modest improvements in efficiency.[11] In contrast, at the dawn of farming, there were large effect innovations to be found, and new domesticates to be discovered. Farmers are thus likely to be more numerous than their foraging neighbors, exacerbating the effects of farmers' sedentary ways. These negative effects of farming on foraging are likely to be obvious to those foragers and thus likely to fuel mutual hostility. James O'Connell (2006) relates a telling anecdote about the effects of a herding culture on the environment of adjoining foragers, and of the hostility and resentment those effects caused.

So the Holocene world was probably one in which intercommunal tension, often spilling over into violence, gradually become more common, and cooperative interactions became less so. This was just one aspect of a revolution in human social life. Farmers are not just sedentary foragers who happen to specialize in gathering rather than hunting. Farming communities were potentially much larger, with all the extra complexities in information gathering and decision making that size alone brings. But in addition to demographic expansion, resources can be accumulated and inherited. Farming societies are not egalitarian. The effects of inheritance were exacerbated by a reduced role for skill. Artisans' skills remained critical in early farming worlds. But Kaplan, Hooper, and Gurven point out that in early farming worlds, extracting resources from the world takes less skill. Breaking ground, weeding, sewing, and harvesting crops are all activities that require less skill than those a forager must command. As a consequence, farming labor can be forced labor. Foragers must be skilled, armed, and mobile, and as a consequence foraging societies do not depend on forced labor. But there have been many social worlds in which farming produce depends on servile labor. Unskilled labor can be compelled because the costs of supervision and coercion can be low compared to their returns. Thus some farming cultures became large, vertically complex, and profoundly inegalitarian.

These inequalities did not just develop between males. Sarah Hrdy (2009) suggests that in the transition to these larger, more hierarchically structured social worlds, female reproductive inequality became much more pronounced. The Pleistocene world of reproductive cooperation was largely one of mutually beneficial bargains between equals. Once equality eroded, allomothering became imposed exploitation by the powerful on the less powerful. With the shift to stratified societies, alloparenting becomes much more costly to the alloparent and depends more heavily on policing and coercion: this is particularly evident with wet-nursing. In stratified societies, female reproductive inequality markedly increases, with subordinate females being forced into quasi sterility. They are forced to delay or limit or abandon their own reproduction to support the reproductive lives of the powerful. There has been plenty of ethnographic discussion of anti-bigman strategies in forager social worlds (for a classic source, see Boehm 1999). Sarah Hrdy points out that the emergence of female inequality in stratified worlds suggests that female leveling strategies were probably equally important in maintaining egalitarian forager social worlds. The breakdown of those strategies is part of the Holocene revolution.

The new social worlds of the Holocene transformed the decision-making environments of individual agents. The informational demands on adaptive choice increased. If Holocene social worlds were indeed larger, Holocene agents needed an expanded database of individual agents, a database that included partial biographical and social assessment, not just recognitional capacity (Dunbar 2001, 2003). Farming worlds are not just larger. Human social worlds include teams, extended families, clans, hunting alliances, and many other stable, functionally important units intermediate between individual agents and the social world as a whole. Some of these were no doubt present in Pleistocene social groups, but Holocene social worlds were larger; they eventually developed top-down mechanisms of command and control, and organized intergroup violence probably played a larger role. So they were more vertically complex. In brief, over the last ten thousand years or so, human social worlds have become larger and more individually heterogeneous. They have more hierarchical structure and more occupational complexity. Agents would have needed more informational resources to make adaptive choices.

Moreover, many of these choices involved collective action, and that poses a further problem. If war was indeed largely a Holocene invention, collective action became more important in the Holocene, not less important. New forms of collective action became important. War is a high-stakes, demanding, collective activity. Moreover, the construction of buildings, walls, and other forms of collective modification of the physical environment became increasingly important throughout the Holocene. Yet the motivation to act collectively is puzzling in many Holocene environments. What is in it for those excluded from wealth and power? If the argument of chapter 5 is right, in foraging social worlds, cooperating in collective activity was typically advantageous for the individuals concerned. But as the social world becomes less egalitarian, many of those involved in collective action will reap a smaller and smaller fraction of cooperation's profit. Why would such agents continue to cooperate as their returns for cooperation decline? These agents were often the majority, and in some early farming-based states, the vast majority of the groups of which they were a part.

Of course, it is not hard to understand participation in collective action in unequal societies once they have developed top-down command-and-control systems backed by coercion. Once influence and position are reinforced by the police, armies, and spies, even for agents who are low in the hierarchy, participation in collective action may well be the least bad option. But such command by coercion systems took many generations to develop (for a good survey of the often slow shift from complex society to early states, see Bogucki 1999). So there must have been a long coexistence of (a) collective action involving many or most members of local groups; (b) significant inequality, and hence significantly unequal returns from collective action; and (c) minimal and inefficient mechanisms of coercive control. The survival—indeed, the elaboration—of cooperation and collective action throughout this period is puzzling. The gradual evolution of elites once farming and husbandry spreads is a triumph of free riding, eventually entrenched by statelike entities. Given the long history of the control of free riding, we need to explain why this trajectory was not derailed by the punishment of incipient elites, or by withdrawing cooperation from them, or both.

One possibility is that even in this transitional period, the less wealthy and influential were better off cooperating, albeit with a reduced share of cooperation's profits. But it is hard to see why that would be true, especially given that elites are necessarily minorities within their social world. Given that the cognitive and cultural machinery for collective action was available, and that elites lacked efficient means of coercion, what prevented the less-wealthy majority from ganging up or excluding the more-wealthy minority? Peter Richerson (with Robert Boyd) and Paul Seabright give an alternative answer. They suggest that features of mind and culture that evolved in Pleistocene-scale worlds were co-opted to support collective action in large-scale worlds, independently of whether such action was adaptive for the agent. Thus Richerson and Boyd (2001) argue that some large-scale social institutions support a cognitive and affective illusion: the illusion of still being part of a tribal world. Paul Seabright (2010) develops a different version of the same idea, one that accords a greater weight to the role of conventions, norms, and fictive kinship systems in motivating prosocial action as the size of human social worlds expands.

Something is missing here. Humans do internalize norms, customs, and systems of categorization, and these surely influence the options we identify and the ways we evaluate them. But the stability of norms and other cultural traits that induce agents to act against their own interests is part of our explanatory target. Consider, for example, Ian Keen's (2006) analysis of Aboriginal foraging cultures. He shows that Aboriginal societies were not markedly economically unequal, but some were markedly sexually unequal. Some were sexual gerontocracies in which older males (and sometimes older females) had far greater access to sexual partners than did their younger counterparts. Older males used their control of ritual knowledge to limit entrée into the adult world. But how could such customs remain stable in the face of the costs they imposed on younger agents? As Brian Haydon points out in commenting on Keen's analysis, we cannot take the kinship and ritual knowledge system of these Aboriginal cultures as explanatorily primitive. Part of what we need to explain is the stability and psychological salience of such systems of belief and evaluation, given the costs that such norms and customs impose, and given that norms and customs are not canalized; people do not always accept the norms and mores of the social worlds into which they are born.

Perhaps intergroup selection plays a role in explaining the prevalence of such norms and customs. Only social entities whose norms, conventions, religions, and kinship systems promoted collective action survived in the dog-eat-dog world of the Holocene. In the Holocene world as Seabright describes it, between-group selection is indeed a powerful force. But it is far from clear that between-group selection could efficiently select for norms, customs, values, and kinship systems that induce cooperation, whether cooperating is in an individual agent's interest or not. Richard Lewontin (1985) observes that selection is a powerful optimizing force when and only when it acts on traits that are free to vary independently of other aspects of an individual's phenotype. Technology and material culture are likely to satisfy this condition. A foraging culture can move from using stone axes they make themselves to steel axes they obtain from trade, while making minimal alterations to the rest of their way of life. But norms, kinship systems, and religious traditions are typically intimately enmeshed with many aspects of a people's way of life; they are not likely to be freely added or discarded, as items of material culture are. So selection will be much less efficient in optimizing them, even in environments in which group selection is powerful (Sterelny 2007).

Clearly, practices of collective action did survive the transitions from foraging to small-scale farming and thence to larger, more vertically complex, inegalitarian, command-and-control societies. Presumably, all these factors played some role in sustaining collective action: group selection in favor of cultural traits that sustain collective action; the co-option of cultural mechanisms that once supported adaptive choices; making the best of a poor hand. All were important. Even so, we have an unsolved chicken-and-egg problem in the transition from the bottom-up organization of collective action in small, relatively egalitarian worlds to the top-down, coercion-backed organization of collective action in larger, inegalitarian social worlds.

Let me finish by summarizing the state of play for the argument as a whole. My overall aim has been to explain human uniqueness: to identify and explain the core, distinctive features of human social life, and some of the cognitive capacities that sustain that life. I have pursued that agenda by developing a picture of human cultural learning built around the apprentice learning model, and of the interplay between informational, ecological, and reproductive cooperation. I have argued that this framework explains the evolution and (limited) stability of behaviorally modern human cultures. Even if I am right in this assessment, the framework clearly leaves open a series of important questions. I have just identified one: perhaps we need something more to explain the stability of distinctively human forms of cooperation across the transition to large-scale, institutionalized, command-and-control societies. There are others. One concerns the generality of the apprentice learning model. I have argued that it generalizes beyond artisan skills, but I have not explored its potential application to many of the domains of human knowledge; in particular, I have not yet used it to explore the vexed question of the evolution of language and symbolic communication. Another concerns the extent and limits of our cognitive plasticity and the relationship between plastic and relatively fixed features of hominin and human psychology. Obviously the general model developed here assumes significant and increasing plasticity. But it does so against a background of more fixed traits (for example, canalized adaptations for social learning). However, I have not developed an explicit and general model of this dynamic, so many questions about human uniqueness and its evolution remain unanswered. But in this book I have identified an explanatory target—the evolution and partial stabilization of behaviorally modern human cultures—and proposed an explanation of that target.

# Notes

## 1 The Challenge of Novelty

1. A point of terminology: I use "hominin" for all the descendants in our branch of the last common ancestor of humans and chimps. I use "human" for the large-brained hominins: *sapiens*, Neanderthals, and their immediate ancestor.

2. For an especially vivid statement of the explanatory problem, see Diamond 1992. For recent reviews of the pattern of hominin history, see Klein and Edgar 2002; Klein 1999; Finlayson 2009.

3. A second point of terminology: I use "social learning" and "cultural learning" interchangeably.

4. By novel, I mean "evolutionarily novel"; I do not mean an individual's first experience of a particular challenge. Obviously, prior experience leading to learning is often essential to adaptive response.

5. For a good recent review of ecological and social transitions in hominin evolution, see Foley and Gamble 2009. One striking feature of the review is the continuing uncertainly about the timing, and even order, of major innovation. Foley and Gamble make their best guesses, but the first domestication of fire and cooking, the origins of humanlike family organization with paternal investment, and the origins of language are all highly conjectural. For example, Foley and Gamble propose that fire was domesticated between one million and eight hundred thousand years ago; Richard Wrangham (2009) places it a million years earlier.

6. The Boxgrove site in England shows evidence of horse hunting around four hundred thousand years ago (Jones 2007); and Foley and Gamble, in their recent review (2009), suggest that large-game hunting was well established and important by then.

7. For example, cycad fruits are grated or pounded with stones and then wrapped in grass parcels that are placed in a stream. This processing leaches the toxins (cycasin and macrozamin), and so the flour can then be baked and the bread eaten (Love 2009).

8. Kristin Hawkes and Jim O'Connell are skeptical about the idea that human life history evolution has been shaped by the demands of cross-generational social learning, but they may not be this skeptical.

9. Stout (2011) argues that despite the apparent similarity over time of the product, Acheulian handaxes were made with increasing skill, and using increasingly complex routines. If he is right, stoneworking skills depended on quite rich and structured cultural input by the late Acheulian, about 0.7 million years ago.

## 2  Accumulating Cognitive Capital

1. This *is* an approximation. Chimps probably inherit some information by social learning. Recent empirical work on social learning increasingly shows that social learning is both widely distributed across animal lineages and important in shaping their lifeways. While some of this information comes from peers and from adults outside the family, some comes from parents (Whiten 2005 and 2011).

2. The existence of these adaptations is not controversial. But their precise identity certainly is. For some contrasting candidates, see Alvard 2003; Csibra and Gergely 2005; Tomasello 1999; Tomasello et al. 2005.

3. The distinction between cue and signal was drawn by Mark Hauser. A cue is an attribute of an agent that others can use as an information source, but does not exist because it is used as an information source. So if one animal sees another take cover, it can use that action as a sign of danger. But the animal taking cover does so not to warn others but to increase its own safety. Likewise gait and posture are cues informing others about age and health, but the old-man shuffle is not produced to inform others of one's increasing decrepitude (Hauser 1996).

4. I discuss a skeptical assessment of this idea in chapter 4. But the controversy rumbles on: its most recent iteration is Gurven and Hill 2010; and Hawkes, O'Connell, and Coxworth 2010.

5. Habitats vary widely, so these numbers only illustrate the delays before efficiency peaks, and the extent of cross-generational material transfer. Few forager societies have been studied quantitatively, and none of the few are ancient.

6. Michael Gurven and his coauthors emphasize this too in discussing the Tsimane (Gurven et al. 2006, 467).

7. There is empirical work attempting to measure the skill demands on cognitive life. But most extant forager groups now use store-bought equipment, so it is difficult to test for the informational demands on making traditional equipment. As a consequence, this empirical work on the cognitive demands of forager skills has been on hunting and gathering itself (see, e.g., Bliege Bird and Bird 2002; Bock 2005).

8. John Shea (2006) points out that the record is likely to contain a lot of practice material, as the acquisition of advanced techniques takes years of practice; we must frequently be seeing the results of children's learning in action.

9. Looking at more recent material, Patricia Crown (2007) makes a similar point. She has demonstrated collaboration between the expert and inexpert in pottery making, both ethnographically and archaeologically, with expert potters often controlling the most difficult parts of the construction process, leaving the less expert (often children) to complete the routine parts. For example, experts lay down the basic design that children then paint in.

## 3   Adapted Individuals, Adapted Environments

1. The human gene pool has undergone significant, selected change in the life of our species, though the specific phenotypic effects of these gene changes are as yet rarely known (Laland, Odling-Smee, and Myles 2010; Nielsen et al. 2007; Varki, Geschwind, and Eichler 2008). Thus gene change may have played some role in the behavioral and cultural changes in the *sapiens* lineage between 200,000 and 50,000 years ago. But if gene change was important to behavior change, it was through a genetic contribution to a gradual, gene-culture coevolutionary process, not a genetically triggered cognitive breakthrough. Or so I argue.

2. A view still defended; see Klein 2008.

3. McBrearty 2001, 92; McBrearty and Tryon 2005, 260–261; Kuhn and Brantingham 2004, 246–248.

4. Indeed, on some views, Neanderthals before their extinction were in the early phase of their own transition to modernity and had *sapiens*-like cognitive horsepower (d'Errico 2003). If so, the gap between the origin of the potential for behaviorally modern lives and its realization may be larger still. Unless Neanderthals and *sapiens* evolved the same capacities by parallel evolution, the potential for behaviorally modern lives would then be in place at the split between the *sapiens* and Neanderthal lineages roughly 500,000 years before present. If d'Errico is right, both branches had the potential to build the materially, ecologically, and symbolically rich cultures of the late Pleistocene and Holocene. But if so, why did those capacities take so long to be expressed?

5. These are not very early, however; none are significantly earlier than about 100,000 years ago (McBrearty 2007).

6. For a somewhat similar suggestion, but in the context of interpersonal interaction, see Kuhn and Stiner 2007a.

7. At least, it has high amplitude when used in large blocks of color. As Peter Hiscock points out to me, it is perfectly possible to use ocher in complex patterns of shapes

and colors, thus giving it the same potential for complex, arbitrary combination as jewelry. He suggests that the real difference might lie in the fact that jewelry can be transferred, and in its potential to last. Especially when used on skin, ocher will have a short shelf-life.

8. Likewise, Steven Mithen's (2005) hypothesis about the role of music is an alternative account of group solidarity that would make self-identifying groups an ancient feature of hominin landscapes.

9. Finlayson is a skeptic, suggesting that accidental dispersal is quite possible. He points to the wide island dispersal of long-tailed macaques as evidence for the power of accidental rafting (Finlayson 2009, 90–92). But a long-tailed macaque weighs between four and eight kilograms, about one-tenth of the weight of an adult human, and is adapted to eating raw food. So a floating tree or tangle of vegetation is proportionally a much larger resource store for a macaque than it would be for a person clinging on. Even so, long-tailed macaques have not reached the Sahul. So I accept the prevailing view that human migration to the Sahul is evidence of technology and planning.

10. Again, we need to be aware that our picture of the earliest Australians may be distorted by biased preservation (Hiscock 2008; Hiscock and O'Connor 2006).

11. A recent and much older date of 3.4 million years is now on the table. This might record a false start, but if it turns out to represent an established and continuing tradition, then the period of conservative technology is even longer (McPherron et al. 2010).

12. April Nowell is so struck by the conservatism of the Acheulian that she argues that *erectus* life history and cognitive capacity conspire against innovation, and that *erectus* lifeways did not select for a more elaborate technology (Nowell 2010b; Nowell and White 2010). But while she is surely right in recognizing fundamental cognitive and developmental differences between erectines and later hominins, indirect evidence shows that they were not as conservative as the stone tool record would suggest.

13. See Henrich 2004; Powell, Shennan, and Thomas 2009; but for a response to Henrich's claim that Tasmania illustrates the effects of these small-world losses, see Read 2006.

14. Henrich and his coworkers claim further that these models show that high-fidelity transmission is not essential for the incremental improvement of skill. While I accept that this work shows the importance of demographic factors to the accumulation of cognitive capital, for reasons I explain in (Sterelny 2011), I do not find this extra claim convincing.

15. The literature also contains hybrid suggestions in which competition strikes the final blow at a species in trouble (Harvarti 2007).

16. See Finlayson and Carrión 2007; for a more consensus view suggesting some significant overlap, see Harvarti 2007. If the recent claim (in Green et al. 2010) that the Neanderthal and sapiens lineages exchanged genes proves to be correct, the two populations must of course have interacted significantly, somewhere, sometime.

17. The ethnographic record supports this idea, and Ambrose interprets the rise of public symbol use in Africa between 100,000 and 50,000 years ago as a response to environmental challenge—building long-distance networks in the face of threat— rather than population growth.

18. Indeed, Bingham's model fits hunting better than it does his original target, within-group enforcement coalitions, for his picture of internal group dynamics is too simple in ways that really matter (Sterelny 2003).

19. These problems would all be exacerbated if, as Stiner and Kuhn argue, Neanderthals did not have a *sapiens*-like division of sexual labor. Once established, the classic forager division of labor has many benefits (more on this in the next chapter). By targeting a wider range of resources, foragers reduce the risks of variable returns. This reduces the vulnerability of the group's reproductive core by encouraging women and children to undertake safer activities. By making the population less dependent on large game, it lowers the human position in the tropic pyramid and hence allows a modest demographic expansion. It maintains a more heterogeneous skill set in the group, buffering it against environmental change. The greater the variety of skills a group has, the more likely it is to be preadapted to change. More generally, the basic model of complementary but cooperative family organization has proved to be adaptive over a wide range of environments. In many, perhaps most, environments, humans foragers depended on a wide range of resources. These are often found at different sites and often require different skills and tools for their exploitation. This combination makes complementary roles adaptive, even when, as in high latitudes, these complementary resources are not foodstuffs. Clothing, shelter, fuel, and fire are crucial female specializations. But while complementary male and female specializations are adaptive over a wide range of environments, Kuhn and Stiner (2006) argue that this division of labor is much more likely to establish in neotropical worlds, where targeting small game and plant-based foods is often profitable. Hence they suggest that this difference in social organization in part explains Neanderthal displacement. D'Errico and Stringer (2011) are skeptical, emphasizing the flexibility of Neanderthal foraging.

20. Birch resin melts at around 340°C but burns at 400°C, so Neanderthal adhesives depended on the precise control of heat (Wadley 2010).

21. Especially, one might suppose, given their physiological head start as über-Inuit.

22. Initially the projectile weapons would likely have been javelins, perhaps using spear-throwers.

## 4   The Human Cooperation Syndrome

1. Unless *H floresiensis* turns out to be a relic habiline, in which case it would show that there has been an otherwise unknown, earlier hominin expansion out of Africa (Groves 2007).

2. These are known as "K-selected" environments. The background idea is that in such environments, the population is at its carrying capacity, so only high-quality offspring can survive to breed. Fertility rates are low, and parents invest heavily in the relatively few offspring they produce.

3. "Males stay, females leave" is a common pattern because strange males are often targets of violence; it is often dangerous in the extreme for males to migrate.

4. The idea that ancient populations were populations of the young depends on skeleton assemblages and estimating age of death from those remains. Hawkes and her colleagues point out that older folks are undercounted because their bones are less likely to survive than younger ones.

5. Kennedy's table 1 shows that in all samples of over 100, more than 10 percent were of older adults (forty years or more) (Kennedy 2003, 555).

6. John Shea (2009, 187) points out that these ancient spears had much larger tips than known, longer-range javelins from the ethnographic record: longer-range weapons are lighter but have higher velocities.

7. The dates of dog domestication are unclear. The fossil evidence suggests that canine domestication was relatively recent, perhaps about 12,000 years ago. Moreover, Australian Aboriginals did not have domestic dogs. But molecular evidence suggests a much deeper date of around 120,000 years (Pennisi 2002).

8. I think it is highly likely that hunting does have reputational effects, and these are sometimes enhanced by the norms and customs surrounding hunting. Thus, apparently, both the Ache and the !Kung have strong norms forbidding boasting by successful hunters, but their success is nonetheless effectively broadcast by the hunter's companions (Boehm 1999; Hawkes and Bird 2002). Such reputational mechanisms are probably secondary modifications of a foraging system that exists because it delivers resources.

9. One might say that the signal is not hunting or hunting effort; it is hunting with a high success rate. And most hunters cannot and do not send that signal. But that would just rephrase the puzzle: we would then need to explain why most hunters attempt to send that signal, knowing that they will fail, and that their failure will expose their lack of star quality. Thanks to Ben Fraser for forcing me to think through these issues more clearly.

10. The more so because Meriam women seem likewise to signal their fitness by collecting industriously and sharing generously. Gurven and Hill (2006) note that

women share as much as men when package size is held constant. They interpret that as a variance-reduction strategy, but it can be fitted into a model in which each sex is under sexual selection.

11. As I noted before, this debate shows no sign of close. For the latest defense of a modified family-provisioning model of male hunting, see Gurven and Hill 2009. Hawkes and her allies reply in Hawkes, O'Connell, and Coxworth 2010, to which Gurven and Hill 2010 is a response. Kramer (2010) reviews the father's role in provisioning (as part of her general review of aid to mothers), concluding that fathers often make important contributions, but their role is quite variable. It is no surprise that in close-to-contemporary foraging societies, the correlations between hunting, sharing, reproduction, and prestige are all noisy and exception ridden.

## 5  Costs and Commitments

1. Classic texts on these issues include Williams 1966; Dawkins 1976; Maynard Smith 1976; Skyrms 1996, 2003; Sober and Wilson 1998. For recent and insightful reviews, see Kerr and Godfrey-Smith 2004; Okasha 2006; West, Griffin, and Gardner 2007; Godfrey-Smith 2009.

2. Thus the emphasis in nativist evolutionary psychology on cheater detection modules and similar policing adaptations. There is remarkably little in this tradition on cognitive specialization for coordination, on modules that support generating cooperative profit, rather than its fair division.

3. The basic structure of a public-goods game involves each agent beginning with an equal stake. Each agent decides whether to assign some fraction of that stake to a common pot. That pot is then increased (usually multiplied by 1.5 or 2), modeling the profit of cooperation, and the expanded pot is then divided equally among the players, independently of their contribution. Bells and whistles can then be added: iterating the game with the same players; adding punishment among the players; allowing third-party onlookers to pay to punish or reward players.

4. For signaling explanations of such phenomena, see Boone 1998, 2000; Sosis et al. 2007.

5. Thus Tomasello and his colleagues show that chimpanzees, unlike humans, rely almost entirely on head direction to track gaze direction; human infants can follow the experimenter's eyes, even when the experimental head is stationary (Tomasello 2009, 76).

6. In an analysis in many ways similar to mine in this chapter, Dan Fessler also emphasizes the significance of investment in relationships; see Fessler and Quintelier, forthcoming.

7. For some preliminary evolutionary explorations of these practices, see Schlegel and Barry 1980; Ludivico and Kurland 1995. For an analysis in terms of group commitment, see Sosis et al. 2007.

8. Somewhat surprisingly, there is some debate about whether this practice is costly. But even setting aside psychological issues, any genital surgery performed in non-sterile conditions must risk serious infection, and we have plenty of evidence of such infections and their biological costs. Both the debate and the costs are discussed in the excellent Mackie 2003; for the alternative view, see Obermeyer 2003.

9. In principle a girl, or more likely her family, might leverage her position by subjecting her to a more extreme version of female genital cutting (likewise, agents might leverage their position in initiation rituals by intensifying or adding to the standard costs). But while important variations exist in the form of female genital cutting (and some are extreme indeed), variation tends to occur across groups rather than within a group; I have not found reports of within-group variation of the kind that would signify variation in intensity of fidelity signals. Jaeger, Caflisch, and Hohlfeld 2009 begins with a useful discussion of the different forms of female genital cutting and of the rate of female genital cutting in the areas of Africa where the practice is common.

## 6  Signals, Cooperation, and Learning

1. This general problem is explored in Cosmides and Tooby 2000; Bacharach and Gambetta 2001; Gambetta 2005; and, especially vividly, in Sperber 1997, 2000, and 2001.

2. Though perhaps not explicitly. There are delicate conceptual and empirical issues here, for some epistemic actions may be hardwired or learned by simple association. The flexible and novel use of epistemic action, and the adept use of cognitive technology, show metarepresentation, but it is not clear just how flexible and novel those uses must be.

3. Clark (2007) gives this insight a distinctive metaphysical spin: he regards some external cognitive resources as genuine constituents of human minds. But we do not have to follow his metaphysical lead to recognize the importance of these scaffolds or their relation to our capacity to think about our own cognitive capacities (Sterelny, 2010).

4. Mode 3 technology is a form of stone toolmaking that involves extensive shaping and preparation of the stone source material before the tools themselves are struck from the core. It apparently requires real skill and emerged roughly 400,000 years ago (Foley and Lahr 2003). For a general analysis of the conditions under which teaching evolves, see Hoppitt et al. 2008; Thornton and Raihani 2008. The crucial thought is that teaching will evolve only when individual, unaided learning is difficult or

dangerous and where some fitness benefit accrues to the teacher, compensating for the cost of teaching. In animal examples—as in meerkats teaching juveniles how to safely catch scorpions—this will normally be an inclusive fitness benefit.

5. It is possible to imagine fanciful situations in which one agent has special access to information but has an interest in the majority view being mistaken, but such scenarios will rarely arise in practice.

6. Chimps and human forager societies known from ethnography live in fission–fusion worlds. That is, the core social group tends to break into small groups that forage together in the day, rejoining in the evening to camp and sleep. It is widely assumed that some form of fission–fusion organization has been typical of hominin social worlds.

7. For example, the resources that an agent brings back from a collecting trip will tell onlookers something about the conditions in the places she has been. Failure to return might hint at unanticipated dangers.

8. See Skyrms 2010 for a modeler's exploration of the conditions under which simple conventional signaling can evolve. As he sees it, the conditions are not restrictive. We do not have to assume error-free signals or perfectly aligned interests.

9. Still less, instead of learning that Andrew wants me to think he likes broccoli.

10. Evidence suggests that quite young children have some sensitivity to varying levels of both epistemic confidence and performance (Harris 2007; Jaswal and Malone 2007).

11. See Tomasello et al. 2005; Tomasello and Carpenter 2007; Tomasello 2008, 2009.

## 7   From Skills to Norms

1. See Dwyer 2006; Hauser 2006; Mikhail 2007; Hauser, Young, and Cushman 2008a,b. The moral grammarians credit the original analogy to John Rawls.

2. Skepticism, however, is increasing. See Cowie 1998; Tomasello 2003, 2008; Devitt 2006; Evans and Levinson 2009.

3. I am perhaps slightly overstating the contrast. Some syntactic, morphological, and phonological variants coincide with ethnic and other social divides, and the folk are able to identify and describe some of these differences.

4. See, e.g., Takezawa, Gummerum, and Keller 2006; Haidt and Bjorklund 2008; Haidt and Kesebir 2010.

5. We might form the belief that a particular sentence is well balanced or awkward, but that belief does not articulate the structural features of the sentence on which understanding depends.

6. Few of our terms for others are normatively neutral: think of "kind," "careless," "unreliable," "cold," "arrogant," and so forth.

7. According to one view, though a highly controversial one, sociopaths do not really understand moral norms, taking them to be conventional-legal norms, because sociopaths lack these emotions.

8. See, e.g., Pizarro 2000; Nichols 2004, 2005; Pizarro et al. 2006; Haidt and Bjorklund 2008; Prinz 2009, 2010. My own views place a greater weight on the role of explicit moral thought and give an explicit and central role to informationally organized developmental environments. But these are just differences in emphasis.

9. Hauser (2006) discusses in this connection the work of Rozin, who shows, for example, that we have no evidence that moral vegetarians have hypersensitive disgust responses, priming their vegetarianism.

10. See Clark 2000 for a view similar to mine on the interaction between tacit and explicit knowledge.

11. Dwyer, if I understand her, thinks this. In particular, she thinks that the distinction between different kinds of norms would be difficult to learn without a big head start. She is impressed by the fact that children have command of the concept of a moral norm, distinguishing it from a merely conventional norm, very early in life. Thus she believes that children must have this concept built in. It is not clear that this datum holds up to further experiment (Kelly et al. 2007; Machery and Mallon 2010). Even if it does, I am less impressed by it than is Dwyer. Tony Scott points out that the norms in question are harm norms, and young children catch on to harm norms, perhaps because violating them results in distinctive emotional signals from the victims (and hence in the onlookers). Moreover, even if parents care about conventional norm violations as intensely as they care about moral norm violations, they offer different justifications and explanations of the two (see Scott 2010, which draws on Smetana 1989, 1999).

## 8   Cooperation and Conflict

1. See, e.g., Jensen, Call, and Tomasello 2007a; Jensen, Call, and Tomasello 2008; Visalberghi and Anderson 2008; for the contrary view, see de Waal and Suchak 2010.

2. For a classic account, see de Waal 2008.

3. Unless gender-, clan-, or age-specific rearing practices entrench systematic differences in the social experience of children within the group.

4. Or to descendant bands formed through some other mechanism, for example, by small parties migrating out to unused habitat.

5. The crucial recent papers are Bowles 2006, 2008, 2009; Bowles and Gintis 2006; Choi and Bowles 2007; Lehmann and Feldman 2008.

6. These results come from public-goods games with punishment, from ultimatum games, and from ultimatum and dictator games with third-party observation (i.e., with observers who can punish unequal divisions).

7. A precise specification of evolutionary altruism is far from easy: it turns out that we can identify an array of suggestions that differ subtly from one another and have importantly different selective signatures. See Kerr and Godfrey-Smith 2004.

8. Thus ultimatum games are transformed if competition takes place among agents who move second. If a proposer can offer to even two agents, and reap the rewards if one accepts, then initial offers are much lower, and low offers are accepted much more often. If the proposer can offer to as many as five competing responding players, then both the offer size and rejection rates shrink to close to those predicted by the *Homo economicus* model (Camerer and Fehr 2006).

9. See also Barclay and Willer 2007, though their studies also showed surprisingly high rates of dishonest signaling: agents acted cooperatively in initial rounds of public-goods games when they knew that they would face contexts of partner choice where their previous history was known. But once chosen, and with no further value to be gained from maintaining their reputation, most failed to act cooperatively.

10. This has sometimes been denied. For example, Don Ross has argued that punishment is often virtually free both because the evolution of human culture has made reputation a real asset (and thus reputation lowering a real threat) and because we have become sensitive to approval and disapproval. So free social sanctions can be substituted for expensive material ones, an idea he supports with a charming example of enforcing round buying at a bar (Ross 2006, 68–69). Ross's scenario is one in which all goes well; the subtly signaled threat of disapproval allows a lapsing free rider to return to good standing with no loss of face. But such social interchanges often come unstuck. The potential seriousness of social and reputational punishment means that if such transactions do go wrong, such punishment produces resentment, anger, social fracture, and counterpunishment. Moreover, our intrinsic (as distinct from instrumental) sensitivity to social sanction is something that we need to explain; it is not an unproblematic explanatory resource.

11. Perhaps the most notable exception is combining horse domestication with pursuit hunting, as in the case of the plains cultures of America. Sadly, zebras and African elephants are not readily domesticated, and so the world was deprived of African pursuit foragers mounted on zebras or elephants, following the Serengeti herds.

# References

Abarbanell, L., and M. Hauser. 2010. Mayan morality: An exploration of permissible harms. *Cognition* 115:207–224.

Allen, J., and J. F. O'Connell. 2008. Getting from Sunda to Sahul. In *Islands of Inquiry: Colonisation, Seafaring, and the Archaeology of Maritime Landscapes*, ed. Geoffrey Clark, Foss Leach, and Sue O'Connor, 31–46. Canberra: ANU E-Press.

Allen-Arave, W., M. Gurven, and K. Hill. 2008. Reciprocal altruism, rather than kin selection, maintains nepotistic food transfers on an Ache reservation. *Evolution and Human Behavior* 29 (5):305–318.

Alperson-Afil, N., D. Richter, and N. Goren-Inbar. 2007. Phantom hearths and the use of fire at Gesher Benot Ya'Aqov, Israel (2007). *PaleoAnthropology* 3:1–15.

Alvard, M. 2003. The adaptive nature of culture. *Evolutionary Anthropology* 12:136–149.

Alvard, M., and D. Nolin. 2002. Rousseau's whale hunt? Coordination among big game hunters. *Current Anthropology* 43 (4):533–559.

Ambrose, S. 2001. Paleolithic technology and human evolution. *Science* 291 (March 2):1748–1753.

Ambrose, S. 2010. Coevolution of composite-tool technology, constructive memory, and language. *Current Anthropology* 51 (Supplement 1):S135–S147.

Apel, J. 2008. Knowledge, know-how, and raw material: The production of late Neolithic flint daggers in Scandinavia. *Journal of Archaeological Methods and Theory* 15:91–111.

Atran, S. 2001. The trouble with memes: Inference versus imitation in cultural creation. *Human Nature* 12 (4):351–381.

Avital, E., and E. Jablonka. 2000. *Animal Traditions: Behavioural Inheritance in Evolution*. Cambridge: Cambridge University Press.

Bacharach, M., and D. Gambetta. 2001. Trust in signs. In *Trust in Society*, vol. II, ed. K. Cook, 148–184. New York: Russell Sage Foundation.

Backwell, L., and F. d'Errico. 2008. Early hominid bone tools from Drimolen, South Africa. *Journal of Archaeological Science* 35 (11):2880–2894.

Bahn, P. 2007. Hunting for clues in the Palaeolithic. *Antiquity* 81:1086–1088.

Baillargeon, R., R. Scott, and Z. He. 2010. False-belief understanding in infants. *Trends in Cognitive Sciences* 14 (3):110–118.

Bamforth, D., and N. Finlay. 2008. Introduction: Archaeological approaches to Lithic production skill and craft learning. *Journal of Archaeological Methods and Theory* 15:1–27.

Barclay, P., and R. Willer. 2007. Partner choice creates competitive altruism in humans. *Proceedings of the Royal Society of London B* 274 (1610):749–753.

Barley, N. 1986. *The Innocent Anthropologist: Notes from a Mud Hut*. London: Penguin.

Bickerton, D. 2002. From protolanguage to language. In *The Speciation of Modern Homo Sapiens*, ed. T. J. Crow, 103–120. Oxford: Oxford University Press.

Bingham, P. 1999. Human uniqueness: A general theory. *Quarterly Review of Biology* 74 (2):133–169.

Bingham, P. 2000. Human evolution and human history: A complete theory. *Evolutionary Anthropology* 9 (6):248–257.

Binmore, K. 2006. Why do people cooperate? *Politics, Philosophy, and Economics* 5:81–96.

Bliege Bird, R., and D. Bird. 2002. Constraints of knowing or constraints of growing? Fishing and collecting by the children of Mer. *Human Nature* 13 (2):239–267.

Bliege Bird, R., D. Bird, B. F. Codding, C. H. Parker, and J. H. Jones. 2008. The "fire stick farming" hypothesis: Australian Aboriginal foraging strategies, biodiversity, and anthropogenic fire mosaics. *Proceedings of the National Academy of Science* 105 (September 30): 14796–14801.

Bliege Bird, R., and E. A. Smith. 2005. Signaling theory, strategic interaction, and symbolic capital. *Current Anthropology* 46 (2):221–248.

Blurton Jones, N., and F. W. Marlowe. 2002. Selection for delayed maturity: Does it take twenty years to learn to hunt and gather? *Human Nature* 13: 199–238.

Bock, J. 2005. What makes a competent adult forager? In *Hunter Gatherer Childhoods: Evolutionary, Developmental, and Cultural Perspectives*, ed. B. S. Hewlett and M. E. Lamb, 109–128. New York: Aldine.

Boehm, C. 1999. *Hierarchy in the Forest*. Cambridge, MA: Harvard University Press.

Boehm, C. 2000. Conflict and the evolution of social control. *Journal of Consciousness Studies* 7 (1–2):79–101.

Bogucki, P. 1999. *The Origins of Human Society*. Oxford: Blackwell.

Boone, J. L. 1998. The evolution of magnanimity. *Human Nature* 9 (1):1–21.

Boone, J. L. 2000. Status signalling, social power, and lineage survival. In *Hierarchies in Action: Cui Bono?*, ed. M. Diehl. Carbondale: Center for Archaeological Investigation.

Bowles, S. 2006. Group competition, reproductive leveling, and the evolution of human altruism. *Science* 314 (5805):1569–1572.

Bowles, S. 2008. Conflict: Altruism's midwife. *Nature* 456:326–327.

Bowles, S. 2009. Did warfare among ancestral hunter-gatherers affect the evolution of human social behaviors? *Science* 324 (5932):1293–1298.

Bowles, S., and H. Gintis. 2003. Origins of human cooperation. In *Genetic and Cultural Evolution of Cooperation*, ed. P. Hammerstein, 429–443. Cambridge, MA: MIT Press.

Bowles, S., and H. Gintis. 2006. The evolutionary basis of collective action. In *The Oxford Handbook of Political Economy*, ed. B. Weingast and D. Wittman, 951–970. Oxford: Oxford University Press.

Boyd, R., H. Gintis, S. Bowles, and P. Richerson. 2005. The evolution of altruistic punishment. In *Moral Sentiments and Material Interests: The Foundations of Cooperation in Economic Life*, ed. H. Gintis, S. Bowles, R. Boyd, and E. Fehr, 215–227. Cambridge, MA: MIT Press.

Boyd, R., and P. Richerson. 1992. Punishment allows the evolution of cooperation (or anything else) in sizable groups. *Ethology and Sociobiology* 13:171–195.

Boyd, R., and P. Richerson. 1996. Why culture is common but cultural evolution is rare. *Proceedings of the British Academy* 88:77–93.

Boysen, S. 1996. "More is less": The elicitation of rule-governed resource distribution in chimpanzees. In *Reaching into Thought: The Minds of the Great Apes*, ed. Anne Russon, Kim Bard, and Sue Taylor Parker, 177–189. Cambridge: Cambridge University Press.

Brown, K., C. Marean, A. Herries, Z. Jacobs, C. Tribolo, D. Braun, D. Roberts, M. Meyer, and J. Bernatchez. 2009. Fire as an engineering tool of early modern humans. *Science* 325:859–862.

Brumm, A., and M. Moore. 2005. Symbolic revolutions and the Australian archaeological record. *Cambridge Archaeological Journal* 15 (2):157–175.

Bureau, P. R. 2001. Abandoning female genital cutting: Prevalence, attitude, and efforts to end the practice. http://www.measurecommunication.org.

Burkart, J., S. B. Hrdy, and C. P. van Schaik. 2009. Cooperative breeding and human cognitive evolution. *Evolutionary Anthropology* 18:175–186.

Byrne, R. 2003. Imitation as behaviour parsing. *Philosophical Transactions of the Royal Society of London, Series B: Biological Sciences* 358:529–536.

Byrne, R. 2004. The manual skills and cognition that lie behind hominid tool use. In *Evolutionary Origins of Great Ape Intelligence*, ed. A. Russon and D. R. Begun, 31–44. Cambridge: Cambridge University Press.

Calcott, B. 2008a. Lineage explanations: Explaining how biological mechanisms change. *British Journal for the Philosophy of Science* 60:51–78.

Calcott, B. 2008b. The other cooperation problem: Generating benefit. *Biology and Philosophy* 23 (2):179–203.

Camerer, C. F., and E. Fehr. 2006. When does "economic man" dominate social behavior? *Science* 311 (5757):47–52.

Carey, J., and D. Judge. 2001. Life span extension in humans is self-reinforcing: A general theory of longevity. *Population and Development Review* 27 (3):411–436.

Carruthers, P. 2006. *The Architecture of the Mind*. New York: Oxford University Press.

Chase, P. 2007. The significance of "acculturation" depends on the meaning of "culture." In *Rethinking the Human Revolution*, ed. Paul Mellars, Katie Boyle, and Ofer Bar-Yosef, 55–66. Cambridge: McDonald Institute for Archaeological Research.

Choi, J.-K., and S. Bowles. 2007. The coevolution of parochial altruism and war. *Science* 318 (5850):636–639.

Christiansen, M., and N. Chater. 2008. Language as shaped by the brain. *Behavioral and Brain Sciences* 31:489–509.

Churchland, P. 1996. The neural representation of the social world. In *Minds and Morals*, ed. L. May, M. Friedman, and A. Clark, 91–108. Cambridge, MA: MIT Press.

Clark, A. 2000. Word and action: Reconciling rules and know-how in moral cognition. In *Moral Epistemology Naturalised*, ed. R. Campbell and B. Hunter, 267–290. Calgary, Alberta: University of Calgary Press.

Clark, A. 2002. Minds, brains, and tools. In *Philosophy of Mental Representation*, ed. Hugh Clapin, 66–90. Oxford: Oxford University Press.

Clark, A. 2008. *Supersizing the Mind: Embodiment, Action, and Cognitive Extension*. Oxford: Oxford University Press.

Clark, A. 2010. *Memento*'s revenge: Objections and replies to the extended mind. In *The Extended Mind*, ed. R. Menary. Cambridge, MA: MIT Press.

Cohen, G. A. 1980. *Karl Marx's Theory of History*. Oxford: Oxford University Press.

Cohen, M. N. 2009. Introduction: Rethinking the origins of agriculture. *Current Anthropology* 50:591–595.

Conard, N. 2006. An overview of the patterns of behavioural change in Africa and Eurasia during the Middle and Late Pleistocene. In *From Tools to Symbols from Early Hominids to Humans*, ed. F. d'Errico and L. Blackwell, 294–332. Johannesburg: Wits University Press.

Conard, N. 2007. Cultural evolution in Africa and Eurasia during the Middle and Late Pleistocene. In *Handbook of Paleoanthropology*, ed. W. Henke and I. Tattersall, 2001–2037. Berlin: Springer.

Conradt, L., and C. List. 2009. Group decisions in humans and animals: A survey. *Philosophical Transaction of the Royal Society B* 364:719–742.

Conradt, L., and T. Roper. 2005. Consensus decision making in animals. *Trends in Ecology and Evolution* 20 (8):449–456.

Coope, G. R. 1994. The response of insect faunas to glacial-interglacial climatic fluctuations. *Philosophical Transactions of the Royal Society of London, Series B: Biological Sciences* 344:19–26.

Cooper, G. 2003. *The Science of the Struggle for Existence*. Cambridge: Cambridge University Press.

Cosmides, L., and J. Tooby. 2000. Consider the sources: The evolution of adaptation for decoupling and metarepresentation. In *Metarepresentation*, ed. D. Sperber, 53–116. New York: Oxford University Press.

Cowie, F. 1998. *What's Within? Nativism Reconsidered*. Oxford: Oxford University Press.

Crown, P. 2007. Life histories of pots and potters: Situating the individual in archaeology. *American Antiquity* 72 (4):677–690.

Csibra, G., and G. Gergely. 2005. Social learning and social cognition: The case for pedagogy. In *Processes of Change in Brain and Cognitive Development*, ed. M. H. Johnson and Y. Munakata. Oxford: Oxford University Press.

Csibra, G., and G. Gergely. 2011. Natural pedagogy as evolutionary adaptation. *Philosophical Transactions of the Royal Society B* 366:1149–1157.

Cushman, F., L. Young, and M. Hauser. 2006. The role of conscious reasoning and intuition in moral judgment: Testing three principles of harm. *Psychological Science* 17 (2):1082–1089.

Damuth, J., and L. Heisler. 1988. Alternative formulations of multilevel selection. *Biology and Philosophy* 3:407–430.

Danchin, E., and G. Luc-Alain. 2004. Public information: From nosy neighbors to cultural evolution. *Science* 305 (July 23).

Dawkins, R. 1976. *The Selfish Gene*. London: Penguin.

Dawkins, R. 2004. Extended phenotype—but not too extended: A reply to Laland, Turner, and Jablonka. *Biology and Philosophy* 19:377–396.

Deacon, T. 1997. *The Symbolic Species: The Co-evolution of Language and the Brain*. New York: W. W. Norton.

de Beaume, S. A. 2004. The invention of technology: Prehistory and cognition. *Current Anthropology* 45 (2):139–162.

Dehaene, S. 1997. *The Number Sense: How the Mind Creates Mathematics*. Oxford: Oxford University Press.

Dennett, D. C. 1983. Intentional systems in cognitive ethology: The "Panglossian paradigm" defended. *Behavioral and Brain Sciences* 6:343–390.

Dennett, D. C. 1988. Out of the armchair and into the field. *Poetics Today* 9:205–221.

Dennett, D. C. 2000. Making tools for thinking. In *Metarepresentation: A Multidisciplinary Perspective*, ed. D. Sperber, 17–29. Oxford: Oxford University Press.

d'Errico, F. 2003. The invisible frontier: A multiple species model for the origin of behavioural modernity. *Evolutionary Anthropology* 12:188–202.

d'Errico, F., C. Henshilwood, M. Vanhaeren, and K. van Niekerk. 2005. Nassarius kraussianus shell beads from Blombos Cave: Evidence for symbolic behaviour in the Middle Stone Age. *Journal of Human Evolution* 48 (1):3–24.

d'Errico, F., and C. Stringer. 2011. Evolution, revolution, or saltation scenario for the emergence of modern cultures? *Philosophical Transactions of the Royal Society, Series B: Biological Sciences* 366:1060–1069.

Dessalles, J.-L. 2007. *Why We Talk: The Evolutionary Origin of Language*. Oxford: Oxford University Press.

Devitt, M. 2006. *Ignorance of Language*. Oxford: Oxford University Press.

de Waal, F. 2008. *Chimpanzee Politics: Power and Sex among Apes*. Baltimore: John Hopkins University Press.

de Waal, F., and M. Suchak. 2010. Prosocial primates: Selfish and unselfish motivations. *Philosophical Transactions of the Royal Society of London, Series B: Biological Sciences* 365 (September):2711–2722.

Diamond, J. 1992. *The Third Chimpanzee: The Evolution and Future of the Human Animal*. New York: HarperCollins.

Dreyfus, H. 1992. *What Computers Still Can't Do: A Critique of Artificial Reason*. Cambridge, MA: MIT Press.

Dunbar, R. 1996. *Grooming, Gossip, and the Origin of Language*. London: Faber.

Dunbar, R. 2001. Brains on two legs: Group size and the evolution of intelligence. In *Tree of Origin*, ed. Franz de Waal, 173–192. Cambridge, MA: Harvard University Press.

Dunbar, R. 2003. The social brain: Mind, language, and society in evolutionary perspective. *Annual Review of Anthropology* 32:163–181.

Dupoux, E., and P. Jacob. 2007. Universal moral grammar: A critical appraisal. *Trends in Cognitive Science* 11 (9):373–378.

Dwyer, S. 2006. How good is the linguistic analogy? In *The Innate Mind*, vol. 2, *Culture and Cognition*, ed. P. Carruthers, S. Laurence, and S. Stich, 237–255. Oxford: Oxford University Press.

Dwyer, S., and M. Hauser. 2008. Dupoux and Jacob's moral instincts: Throwing out the baby, the bathwater, and the bathtub. *Trends in Cognitive Science* 12 (1):1–2.

Edgerton, R. B. 1992. *Sick Societies: Challenging the Myth of Primitive Harmony*. New York: Free Press.

Ennion, E. A. R., and N. Tinbergen. 1967. *Tracks*. Oxford: Clarendon Press.

Enquist, M., and S. Ghirlanda. 2007. Evolution of social learning does not explain the origin of human cumulative culture. *Journal of Theoretical Biology* 246:129–135.

Evans, N., and S. Levinson. 2009. The myth of language universals: Language diversity and its importance for cognitive science. *Behavioral and Brain Sciences* 32: 429–492.

Everett, D. L. 2005. Cultural constraints on grammar and cognition in Pirahã: Another look at the design features of human language. *Current Anthropology* 46 (4):621–646.

Fehr, E., and U. Fischbacher. 2003. The nature of human altruism. *Nature* 425:785–791.

Fehr, E., and U. Fischbacher. 2004. Social norms and human cooperation. *Trends in Cognitive Sciences* 8 (4):185–189.

Fessler, D., and K. Quintelier. Forthcoming. Suicide bombers, weddings, and prison tattoos: An evolutionary perspective on subjective commitment and objective commitment. In *Signaling, Commitment, and Emotion*, ed. B. Calcott, B. Fraser, R. Joyce, and K. Sterelny. Cambridge, MA: MIT Press.

Finlayson, C. 2005. Biogeography and evolution of the genus Homo. *Trends in Ecology and Evolution* 20 (August):457–463.

Finlayson, C. 2009. *The Humans Who Went Extinct: Why Neanderthals Died Out and We Survived*. New York: Oxford University Press.

Finlayson, C., and J. Carrión. 2007. Rapid ecological turnover and its impact on Neanderthal and other human populations. *Trends in Ecology and Evolution* 22 (April):213–222.

Flinn, M., D. C. Geary, and C. Ward. 2005. Ecological dominance, social competition, and coevolutionary arms races: Why humans evolved extraordinary intelligence. *Evolution and Human Behavior* 26:10–46.

Flynn, E. 2008. Investigating children as cultural magnets: Do young children transmit redundant information along diffusion chains? *Philosophical Transactions of the Royal Society of London, Series B: Biological Sciences* 363:3541–3551.

Foley, R. 1995. *Humans before Humanity: An Evolutionary Perspective*. Oxford: Blackwell.

Foley, R. 2002. Adaptive radiations and dispersals in hominin evolutionary ecology. *Evolutionary Anthropology* 11 (S1):32–37.

Foley, R., and C. Gamble. 2009. The ecology of social transitions in human evolution. *Philosophical Transactions of the Royal Society of London, Series B: Biological Sciences* 364:3267–3279.

Foley, R., and M. Lahr. 2003. On stony ground: Lithic technology, human evolution, and the emergence of culture. *Evolutionary Anthropology* 12:109–122.

Foley, R., and M. Lahr. 2011. The evolution and diversity of cultures. *Philosophical Transactions of the Royal Society, Series B: Biological Sciences* 366:1080–1089.

Foley, R., and P. C. Lee. 1989. Finite social space, evolutionary pathways, and reconstructing hominid behavior. *Science* 243:901–906.

Frank, M. C., D. L. Everett, E. Fedorenko, and E. Gibson. 2008. Number as a cognitive technology: Evidence from Pirahã language and cognition. *Cognition* 108 (3):819–824.

Frank, R. 1988. *Passion within Reason: The Strategic Role of the Emotions*. New York: W. W. Norton.

Frank, R. 2001. Cooperation through emotional commitment. In *Evolution and the Capacity for Commitment*, ed. R. Nesse, 57–77. New York: Russell Sage Foundation.

Fraser, G. M. 2007. *Quartered Safe Out Here: A Recollection of the War in Burma*. London: Penguin.

Gächter, S., and B. Herrmann. 2009. Reciprocity, culture, and human cooperation: Previous insights and a new cross-cultural experiment. *Philosophical Transactions of the Royal Society of London, Series B: Biological Sciences* 364 (1518):791–806.

Gächter, S., B. Herrmann, and C. Thoni. 2010. Culture and cooperation. *Philosophical Transactions of the Royal Society of London, Series B: Biological Sciences* 365: 2651–2661.

Gaffin, D. 1995. The production of emotion and social control: Taunting, anger, and the Rukka in the Faeroe Islands. *Ethos* 23 (2):149–172.

Galef, B., and K. Laland. 2005. Social learning in animals: Empirical studies and theoretical models. *Bioscience* 55 (6):489–499.

Gambetta, D. 2005. Deceptive mimicry in humans. In *Perspectives on Imitation from Neuroscience to Social Science*, vol. 2, ed. S. Hurley and N. Chater, 221–241. Cambridge, MA: MIT Press.

Gergely, G., and G. Csibra. 2005. The social construction of the cultural mind. *Interaction Studies: Social Behaviour and Communication in Biological and Artificial Systems* 6:465–481.

Gergely, G., and G. Csibra. 2006. Sylvia's recipe: The role of imitation and pedagogy in the transmission of cultural knowledge. In *Roots of Human Society: Culture, Cognition, and Human Interaction*, ed. N. J. Enfield and S. C. Levenson, 229–255. Oxford: Berg.

Gergely, G., K. Egyed, and I. Kiraly. 2007. On pedagogy. *Developmental Science* 10 (1):139–146.

Gil-White, F. 2005. Common misunderstandings of memes (and genes): The promise and the limits of the genetic analogy to cultural transmission processes. In *Perspectives on Imitation: From Mirror Neurons to Memes*, ed. S. Hurley and N. Chater. Cambridge, MA: MIT Press.

Gilligan, I. 2007. Neanderthal extinction and modern human behaviour: The role of climate change and clothing. *World Archaeology* 39 (4):499–514.

Gintis, H. 2006. Behavioral ethics meets natural justice. *Politics, Philosophy, and Economics* 5 (1):5–32.

Gintis, H. 2008. Punishment and co-operation. *Science* 319:1345–1346.

Gintis, H., S. Bowles, R. Boyd, and E. Fehr, eds. 2005. *Moral Sentiments and Material Interests*. Cambridge, MA: MIT Press.

Godfrey-Smith, P. 2009. *Darwinian Populations and Natural Selection*. Oxford: Oxford University Press.

Goren-Inbar, N. 2011. Culture and cognition in the Acheulian industry: A case study from Gesher Benot Ya'aqov. *Philosophical Transactions of the Royal Society, Series B: Biological Sciences* 366:1038–1049.

Gopnik, A. In press. Causality. In *The Oxford Handbook of Developmental Psychology*, ed. P. Zelazo. New York: Oxford University Press.

Grafen, A. 1990. Biological signals as handicaps. *Journal of Theoretical Biology* 144:517–546.

Graves, R. 1929. *Good-Bye to All That*. London.

Green, R. E., J. Krause, A. W. Briggs, T. Maricic, et al. 2010. A draft sequence of the Neandertal genome. *Science* 328 (5979):710–722.

Grimm, L. 2000. Apprentice flintknapping: Relating material culture and social practice in the Upper Palaeolithic. In *Children and Material Culture*, ed. Joanna Safaer Derevenski, 53–71. New York: Routledge.

Groves, C. 2007. The Homo floresiensis controversy. *Hayati Journal of Biosciences* 14 (4):123–126.

Gurven, M. 2004. To give and to give not: The behavioral ecology of human food transfers. *Behavioral and Brain Sciences* 27:543–583.

Gurven, M., and K. Hill. 2006. Hunting as subsistence and mating effort? A reevaluation of "Man the Hunter," the sexual division of labor, and the evolution of the nuclear family. In *IUSSP Seminar on Male Life History*. Giessen, Germany.

Gurven, M., and K. Hill. 2009. Why do men hunt? A reevaluation of "Man the Hunter" and the sexual division of labor. *Current Anthropology* 50 (1):51–74.

Gurven, M., and K. Hill. 2010. Moving beyond stereotypes of men's foraging goals. *Current Anthropology* 51 (2):265–267.

Gurven, M., H. Kaplan, and M. Gutierrez. 2006. How long does it take to become a proficient hunter? Implications for the evolution of extended development and long life span. *Journal of Human Evolution* 51 (5):451–470.

Guthrie, R. D. 2005. *The Nature of Paleolithic Art*. Chicago: University of Chicago Press.

Haagen, C. 1994. *Bush Toys: Aboriginal Children at Play*. Canberra: Aboriginal Studies Press.

Habgood, P., and N. Franklin. 2008. The revolution that didn't arrive: A review of Pleistocene Sahul. *Journal of Human Evolution* 55:187–222.

Hagen, E. H., and G. A. Bryant. 2003. Music and dance as a coalition signaling system. *Human Nature* 14 (1):21–51.

Haidt, J., and F. Bjorklund. 2008. Social intuitionists answer six questions about moral psychology. In *Moral Psychology*, vol. 2, *The Cognitive Science of Morality: Intuition and Diversity*, ed. W. Sinnott Armstrong, 181–217. Cambridge, MA: MIT Press.

Haidt, J., and S. Kesebir. 2010. Morality. In *Handbook of Social Psychology*, 5th ed., ed. S. Fiske, D. Gilbert, and G. Lindzey, 797–832. Hoboken, NJ: Wiley.

Hammerstein, P., ed. 2003. *Genetic and Cultural Evolution of Cooperation*. Cambridge, MA: MIT Press.

Harris, P. L. 2007. Trust. *Developmental Science* 10 (1):135–138.

Harvarti, K. 2007. Neanderthals and their contemporaries. In *Handbook of Paleoanthropology*, ed. Winfried Henke and Ian Tattersall, 1717–1748. Berlin: Springer.

Hauser, M. 1996. *The Evolution of Communication*. Cambridge, MA: MIT Press.

Hauser, M. 2006. *Moral Minds: How Nature Designed Our Universal Sense of Right and Wrong*. New York: HarperCollins.

Hauser, M., L. Young, and F. Cushman. 2008a. On misreading the linguistic analogy: Response to Jesse Prinz and Ron Mallon. *Moral Psychology and Biology*, ed. W. Sinnott Armstrong, 171–180. Oxford: Oxford University Press.

Hauser, M., L. Young, and F. Cushman. 2008b. Reviving Rawls's linguistic analogy. *Moral Psychology and Biology*, ed. W. Sinnott Armstrong, 107–144. Oxford: Oxford University Press.

Hawkes, K. 1991. Showing-off: Tests of another hypothesis about men's foraging goals. *Ethology and Sociobiology* 11:29–54.

Hawkes, K. 1994. The grandmother effect. *Nature* 428:128–129.

Hawkes, K. 2003. Grandmothers and the evolution of human longevity. *American Journal of Human Biology* 15 (3):380–400.

Hawkes, K., and R. Bird. 2002. Showing off, handicap signaling, and the evolution of men's work. *Evolutionary Anthropology* 11 (1):58–67.

Hawkes, K., J. F. O'Connell, N. G. Blurton Jones, H. Alvarez, and E. Charnov. 1998. Grandmothering, menopause, and the evolution of human life histories. *Proceedings of the National Academy of Sciences of the United States of America* 95:1336–1339.

Hawkes, K., J. F. O'Connell, and J. E. Coxworth. 2010. Family provisioning is not the only reason men hunt. *Current Anthropology* 51 (2):259–264.

Henrich, J. 2004. Demography and cultural evolution: Why adaptive cultural processes produced maladaptive losses in Tasmania. *American Antiquity* 69 (2): 197–221.

Henrich, J., and R. Boyd. 2002. On modeling cognition and culture: Why cultural evolution does not require replication of representations. *Journal of Cognition and Culture* 2:87–112.

Henrich, J., R. Boyd, S. Bowles, C. Camerer, E. Fehr, H. Gintis, R. McElreath, M. Alvard, A. Barr, and J. Ensminger. 2005. "Economic man" in cross-cultural perspective: Behavioral experiments in 15 small-scale societies. *Behavioral and Brain Sciences* 28 (6):795–814.

Henrich, J., R. Boyd, and P. Richerson. 2008. Five misunderstandings about cultural evolution. *Human Nature* 19 (2):119–137.

Henrich, J., and F. Gil-White. 2001. The evolution of prestige: Freely conferred deference as a mechanism for enhancing the benefits of cultural transmission. *Evolution and Human Behavior* 22:165–196.

Henshilwood, C., and C. Marean. 2003. The origin of modern behavior. *Current Anthropology* 44 (5):627–651. [Includes peer commentary and author's responses.]

Herrmann, E., J. Call, M. V. Hernandez-Lloreda, B. Hare, and M. Tomasello. 2007. Humans have evolved specialized skills of social cognition: The cultural intelligence hypothesis. *Science* 317 (September 7):1360–1366.

Hewlett, B., H. Fouts, A. Boyette, and B. Hewlett. 2011. Social learning among Congo Basin hunter-gatherers. *Philosophical Transactions of the Royal Society, Series B: Biological Sciences* 366:1168–1178.

Heyes, C. 2011. Automatic imitation. *Psychological Bulletin* 137 (3):463–483.

Hill, K., and H. Kaplan. 1999. Life history traits in humans: Theory and empirical studies. *Annual Review of Anthropology* 28:397–430.

Hiscock, P. 2008. *Archaeology of Ancient Australia*. London: Routledge.

Hiscock, P., and S. O'Connor. 2006. An Australian perspective on modern behaviour and artefact assemblages. *Before Farming* 2:1–10.

Hiscock, P., A. Turq, J.-P. Faivre, and L. Bourguignon. 2009. Quina procurement and tool production. In *Lithic Materials and Paleolithic Societies*, ed. B. Adams and B. Blades, 232–246. Oxford: Blackwell.

Holzhaider, J. C., G. R. Hunt, and R. D. Gray. 2010a. Social learning in New Caledonian crows. *Learning and Behaviour* 38:206–219.

Holzhaider, J.C., G. R. Hunt, and R. D. Gray. 2010b. The development of pandanus tool manufacture in wild New Caledonian crows. *Behaviour* 147:553–586.

Hoppitt, W. J. E., G. R. Brown, R. Kendal, L. Rendell, A. Thornton, M. M. Webster, and K. N. Laland. 2008. Lessons from animal teaching. *Trends in Ecology and Evolution* 23:486–493.

Hrdy, S. B. 1999. *Mother Nature: A History of Mothers, Infants, and Natural Selection*. New York: Pantheon Books.

Hrdy, S. B. 2005. Evolutionary context of development: The cooperative breeding model. In *Attachment and Bonding: A New Synthesis*, ed. C. S. Carter, L. Ahnert, K. E. Grossman, S. B. Hrdy, M. E. Lamb, S. W. Porges, and N. Sachser, 9–32. Cambridge, MA: MIT Press.

Hrdy, S. B. 2009. *Mothers and Others: The Evolutionary Origins of Mutual Understanding*. Cambridge, MA: Harvard University Press.

Huebner, B., S. Dwyer, and M. Hauser. 2009: The role of emotion in moral psychology. *Trends in Cognitive Science* 13:1–6.

Humphrey, N. 1976. The social function of intellect. In *Growing Points in Ethology*, ed. P. P. G. Bateson and R. A. Hinde, 303–317. Cambridge: Cambridge University Press.

Ikandal, D., and C. Packer. 2008. Ritual vs. retaliatory killing of African lions in the Ngorongoro Conservation Area, Tanzania. *Endangered Species Research* 6:67–74.

Jablonka, E., and M. Lamb. 2005. *Evolution in Four Dimensions*. Cambridge, MA: MIT Press.

Jaeger, F., M. Caflisch, and P. Hohlfeld. 2009. Female genital mutilation and its prevention: A challenge for paediatricians. *European Journal of Pediatrics* 168 (1):27–33.

Jaswal, V. K., and L. S. Malone. 2007. Turning believers into skeptics: 3-year-olds' sensitivity to cues to speaker credibility. *Journal of Cognition and Development* 8 (3):263–283.

Jensen, K., J. Call, and M. Tomasello. 2007a. Chimpanzees are rational maximizers in an ultimatum game. *Science* 318: 107–109.

Jensen, K., J. Call, and M. Tomasello. 2007b. Chimpanzees are vengeful but not spiteful. *Proceedings of the National Academy of Sciences, USA* 104 (32):13046–13050.

Jensen, K., J. Call, and M. Tomasello. 2008. Response. *Science* 319:283.

Jones, M. 2007. *Feast: Why Humans Share Food*. Oxford: Oxford University Press.

Joyce, R. 2006. *Evolution of Morality*. Cambridge, MA: MIT Press.

Kaplan, H., M. Gurven, K. Hill, and M. Hurtado. 2005. The natural history of human food sharing and cooperation: A review and a new multi-individual approach to the negotiation of norms. In *The Moral Sentiments and Material Interests: The Foundations of Cooperation in Economic Life*, ed. S. Bowles, R. Boyd, E. Fehr, and H. Gintis. Cambridge, MA: MIT Press.

Kaplan, H., K. Hill, J. Lancaster, and M. Hurtado. 2000. A theory of human life history evolution: Diet, intelligence, and longevity. *Evolutionary Anthropology* 9 (4):156–185.

Kaplan, H., P. Hooper, and M. Gurven. 2009. The evolutionary and ecological roots of human social organization. *Philosophical Transactions of the Royal Society of London, Series B: Biological Sciences* 364:3289–3299.

Kauffman, S. 1995. *At Home in the Universe: The Search for Laws of Self-Organization and Complexity*. Oxford: Oxford University Press.

Keegan, J. 1983. *The Face of Battle*. London: Penguin.

Keen, I. 2006. Constraints on the development of enduring inequalities in Late Holocene Australia. *Current Anthropology* 47 (1):7–38.

Kelly, R. 2005. The evolution of lethal intergroup violence. *Proceedings of the National Academy of Sciences of the United States of America* 102 (43):15294–15298.

Kelly, D., and S. Stich, et al. 2007. Harm, affect, and the moral/conventional distinction. *Mind and Language* 22:117–131.

Kennedy, G. 2003. Palaeolithic grandmothers? Life history theory and early Homo. *Journal of the Royal Anthropological Institute* 8:549–572.

Kennedy, G. 2005. From the ape's dilemma to the weanling's dilemma: Early weaning and its evolutionary context. *Journal of Human Evolution* 48:123–145.

Kerr, B., and P. Godfrey-Smith. 2004. What is altruism? *Trends in Ecology and Evolution* 19 (3):135–140.

Klein, R. 1999. *The Human Career: Human Biological and Cultural Origins*. 2nd ed. Chicago: University of Chicago Press.

Klein, R. 2008. Out of Africa and the evolution of human behavior. *Evolutionary Anthropology* 17:267–281.

Klein, R. 2009. Darwin and the recent African origin of modern humans. *Proceedings of the National Academy of Sciences of the United States of America* 106 (38):16007–16009.

Klein, R., and B. Edgar. 2002. *The Dawn of Human Culture*. New York: Wiley.

Kramer, K. L. 2010. Cooperative breeding and its significance to the demographic success of humans. *Annual Review of Anthropology* 39: 417–436.

Krebs, J., and R. Dawkins. 1984. Animal signals, mind-reading, and manipulation. In *Behavioural Ecology: An Evolutionary Approach*, ed. J. R. Krebs and N. B. Davies, 380–402. Oxford: Blackwell Scientific.

Kuhn, S., and P. J. Brantingham. 2004. The Early Upper Paleolithic and the origins of modern human behavior. In *The Early Upper Paleolithic beyond Western Europe*, ed. P. J. Brantingham, S. Kuhn, and K. W. Kerry, 242–248. Berkeley: University of California Press.

Kuhn, S., and M. C. Stiner. 2006. What's a mother to do? *Current Anthropology* 47 (6):953–980.

Kuhn, S., and M. C. Stiner. 2007a. Body ornamentation as information technology: Towards an understanding of the significance of early beads. In *Rethinking the Human Revolution*, ed. Paul Mellars, Katie Boyle, Ofer Bar-Yosef, and Chris Stringer, 45–54. Cambridge: McDonald Institute for Archaeological Research.

Kuhn, S., and M. C. Stiner. 2007b. Palaeolithic ornaments: Implications for cognition, demography, and identity. *Diogenes* 214:40–48.

Kuhn, S., M. C. Stiner, D. S. Reese, and E. Güleç. 2001. Ornaments of the earliest Upper Paleolithic: New insights from the Levant. *Proceedings of the National Academy of Sciences of the United States of America* 98 (13):7641–7646.

Laland, K. 2001. Imitation, social learning, and preparedness as mechanisms of bounded rationality. In *Bounded Rationality: The Adaptive Toolbox*, ed. G. Gigerenzer and R. Selten, 233–248. Cambridge, MA: MIT Press.

Laland, K. 2007. Niche construction, human behavioural ecology, and evolutionary psychology. In *Oxford Handbook of Evolutionary Psychology*, ed. R. Dunbar and L. Barrett, 35–48. Oxford: Oxford University Press.

Laland, K., and G. Brown. 2002. *Sense and Nonsense: Evolutionary Perspectives on Human Behaviour*. Oxford: Oxford University Press.

Laland, K., and B. Galef, eds. 2009. *The Question of Animal Culture*. Cambridge, MA: Harvard University Press.

Laland, K., and W. Hoppitt. 2003. Do animals have culture? *Evolutionary Anthropology* 12:150–159.

Laland, K., J. Odling-Smee, and M. W. Feldman. 2000. Niche construction, biological evolution, and cultural change. *Behavioral and Brain Sciences* 23:131–175.

Laland, K., J. Odling-Smee, and S. Myles. 2010. How culture shaped the human genome: Bringing genetics and the human sciences together. *Nature Reviews: Genetics* 11:137–148.

Lave, J. 1996. Teaching, as learning, in practice. *Mind, Culture, and Activity* 3 (3): 149–164.

Lehmann, L., and M. Feldman. 2008. War and the evolution of belligerence and bravery. *Proceedings: Biological Sciences* 275 (1653):2877–2885.

Lewontin, R. C. 1985. Adaptation. In R. Levins and R. Lewontin, *The Dialectical Biologist*. Cambridge, MA: Harvard University Press.

Lewontin, R. C. 2000. *The Triple Helix*. Cambridge, MA: Harvard University Press.

Liebenberg, L. 1990. *The Art of Tracking and the Origin of Science*. Claremount, South Africa: David Philip.

Liebenberg, L. 2008. The relevance of persistence hunting to human evolution. *Journal of Human Evolution* 55 (6):1156–1159.

Lieberman, P. 1998. *Eve Spoke: Human Language and Human Evolution*. New York: W. W. Norton.

List, C. 2004. Democracy in animal groups: A political science perspective. *Trends in Ecology and Evolution* 19 (4):168–169.

Love, J. R. B. 2009. *Stone Age Bushmen of Today.* Ed. D. M. Welch. Australian Aboriginal Culture Series No. 6. Virginia, NT: D. M. Welch. (Originally published in 1936.)

Ludivico, L., and J. A. Kurland. 1995. Symbolic or not-so-symbolic wounds: The behavioral ecology of human scarification. *Ethology and Sociobiology* 16:155–172.

Lycett, S., and J. Gowlett. 2008. On questions surrounding the Acheulean "tradition." *World Archaeology* 40 (3):295–315.

MacDonald, K. 2007. Cross-cultural comparison of learning in human hunting: Implications for life history evolution. *Human Nature* 18:386–402.

MacDonald, L. 1993. *Somme.* London: Penguin.

MacDonald, L. 1994. *They Called It Passchendaele.* London: Penguin.

Mace, R. 2009. On becoming modern. *Science* 324 (5932):1280–1281.

Machery, E., and R. Mallon. 2010. The evolution of morality. In *The Oxford Handbook of Moral Psychology*, ed. J. M. Doris, F. Cushman et al., 1–13. Oxford: Oxford University Press.

Mackie, G. 1996. Ending footbinding and infibulation: A convention account. *American Sociological Review* 61 (6):999–1017.

Mackie, G. 2003. Female genital cutting: A harmless practice? *Medical Anthropology Quarterly* 17 (2):135–158.

Mameli, M. 2004. Nongenetic selection and nongenetic inheritance. *British Journal for the Philosophy of Science* 55 (1):35–71.

Marean, C., M. Bar-Matthews, J. Bernatchez, E. Fisher, P. Goldberg, A. I. Herries, Z. Jacobs, A. Jerardino, P. Karkanas, T. Minichillo, P. J. Nilssen, E. Thompson, I. Watts, and H. M. Williams. 2007. Early human use of marine resources and pigment in South Africa during the Middle Pleistocene. *Nature* 449:905–908.

Marlowe, F. W. 2005. Hunter-gatherers and human evolution. *Evolutionary Anthropology* 14:54–67.

Marlowe, F. W. 2007. Hunting and gathering: The human sexual division of foraging labor. *Cross-Cultural Research* 41:170–194.

Maynard Smith, J. 1976. Group selection. *Quarterly Review of Biology* 51:277–283.

McBrearty, S. 2001. The Middle Pleistocene of East Africa. In *Human Roots: Africa and Asia in the Middle Pleistocene*, ed. L. Barham and K. Robson-Brown, 81–97. Bristol: Western Academic.

McBrearty, S. 2007. Down with the revolution. In *Rethinking the Human Revolution: New Behavioural and Biological Perspectives on the Origin and Dispersal of Modern*

*Humans*, ed. P. Mellars, K. Boyle, O. Bar-Yosef, and C. Stringer, 133–151. Cambridge: McDonald Institute Archaeological Publications.

McBrearty, S., and A. Brooks. 2000. The revolution that wasn't: A new interpretation of the origin of modern human behavior. *Journal of Human Evolution* 39 (5):453–563.

McBrearty, S., and C. Stringer. 2007. The coast in colour. *Nature* 449:793–794.

McBrearty, S., and C. Tryon. 2005. From Acheulean to Middle Stone Age in the Kapthurin Formation, Kenya. In *Transitions before the Transition: Evolution and Stability in the Middle Paleolithic and Middle Stone Age*, ed. E. Hovers and S. Kuhn, 257–277. Dordrecht: Springer.

McNabb, J. 2005. Reply. *Current Anthropology* 46 (3):460–463.

McNabb, J., F. Binyon, and L. Hazelwood. 2004. The large cutting tools from the South African Acheulean and the question of social traditions. *Current Anthropology* 45 (5):653–677.

McPherron, S., Z. Alemseged, C. Marean, J. Wynn, D. Reed, D. Geraads, R. Bobe, and H. Béarat. 2010. Evidence for stone-tool-assisted consumption of animal tissues before 3.39 million years ago at Dikika, Ethiopia. *Nature* 466:857–860.

Mellars, P. 2005. The impossible coincidence: A single-species model for the origins of modern human behavior in Europe. *Evolutionary Anthropology* 14:12–27.

Mercier, H., and D. Sperber. 2011. Why do humans reason? Arguments for an argumentative theory. *Behavioral and Brain Sciences* 35:57–74.

Mesoudi, A., and A. Whiten. 2008. The multiple roles of cultural transmission experiments in understanding human cultural evolution. *Philosophical Transactions of the Royal Society of London, Series B: Biological Sciences* 363:3489–3501.

Mikhail, J. 2007. Universal moral grammar: Theory, evidence, and the future. *Trends in Cognitive Science* 11(4):143–152.

Miller, G. F. 1997. Sexual selection for moral virtues. *Quarterly Review of Biology* 82 (2):97–125.

Mithen, S. 2005. *The Singing Neanderthals: The Origins of Music, Language, Mind, and Body*. London: Weidenfeld & Nicolson.

Montefiore, S. 2005. *Stalin: The Court of the Red Tsar*. London: Orion Books.

Morrison, R. 1981. *A Field Guide to the Tracks and Traces of Australian Animals*. Adelaide: Rigby.

Nesse, R. 2001. *Evolution and the Capacity for Commitment*. New York: Russell Sage.

Nichols, S. 2004. *Sentimental Rules: On the Natural Foundations of Moral Judgment*. New York: Oxford University Press.

Nichols, S. 2005. Innateness and moral psychology. In *The Innate Mind: Structure and Content*, ed. P. Carruthers and S. Laurence, 353–370. New York: Oxford University Press.

Nielsen, R., I. Hellmann, M. Hubisz, C. Bustamante, and A. G. Clark. 2007. Recent and ongoing selection in the human genome. *Nature Reviews: Genetics* 8:857–868.

Nisbett, R. E., and D. Cohen. 1996. *Culture of Honour*. Boulder, CO: Westview Press.

Noe, R. 2006. Cooperation experiments: Coordination through communication versus acting apart together. *Animal Behaviour* 71:1–18.

Nowak, M., and K. Sigmund. 2005. Evolution of indirect reciprocity. *Nature* 437 (7063):1291–1298.

Nowell, A. 2010a. Defining behavioural modernity in the context of Neandertal and anatomically modern human populations. *Annual Review of Anthropology* 39:437–452.

Nowell, A. 2010b. Working memory and the speed of life. *Current Anthropology* 51 (Supplement 1):S121–S133.

Nowell, A. Forthcoming. Cognition, behavioral modernity, and the archaeological record of the Middle and Early Upper Paleolithic. In *Evolution of Mind, Brain, and Culture*, ed. Gary Hatfield. Philadelphia: Penn Museum Press.

Nowell, A., and M. White. 2010. Growing up in the Middle Pleistocene. In *Stone Tools and the Evolution of Human Cognition*, ed. A. Nowell and I. Davidson, 67–82. Boulder, CO: University of Colorado Press.

O'Connell, J. F. 2006. How did modern humans displace Neanderthals? Insights from hunter-gatherer ethnography and archaeology. In *When Neanderthals and Modern Humans Met*, ed. N. Conard, 43–64. Tübingen: Kerns.

O'Connell, J. F., and J. Allen. 2007. Pre-LGM Sahul (Pleistocene Australia–New Guinea) and the archaeology of early modern humans. In *Rethinking the Human Revolution*, ed. P. Mellars, K. Boyle, O. Bar-Yosef, and C. Stringer, 395–410. Cambridge: McDonald Institute for Archaeological Research.

O'Connell, J. F., K. Hawkes, and N. Blurton Jones. 1999. Grandmothering and the evolution of Homo erectus. *Journal of Human Evolution* 36 (5):461–485.

Obermeyer, C. M. 2003. The health consequences of female circumcision: Science, advocacy, and standards of evidence. *Medical Anthropology Quarterly* 17 (3):394–411.

Odling-Smee, J., K. Laland, and M. Feldman. 2003. *Niche Construction: The Neglected Process in Evolution*. Princeton: Princeton University Press.

Ofek, H. 2001. *Second Nature: Economic Origins of Human Evolution*. Cambridge: Cambridge University Press.

Okasha, S. 2006. *Evolution and the Units of Selection*. Oxford: Oxford University Press.

Ostrom, E. 1998. A behavioral approach to the rational choice theory of collective action. *American Political Science Review* 92 (1):1–22.

Pennisi, E. 2002. Canine evolution: A shaggy dog history. *Science* 298 (5598):1540–1542.

Pinker, S. 1994. *The Language Instinct: How the Mind Creates Language.* New York: William Morrow.

Pinker, S. 1997. *How the Mind Works.* New York: W. W. Norton.

Pinker, S. 2007. *The Stuff of Thought: Language as a Window into Human Nature.* New York: Penguin.

Pizarro, D. 2000. Nothing more than feelings? The role of emotions in moral judgement. *Journal for the Theory of Social Behaviour* 30 (4):355–375.

Pizarro, D., B. Detweiler-Bedell, and P. Bloom. 2006. The creativity of everyday moral reasoning: Empathy, disgust, and moral persuasion. In *Creativity and Reason in Cognitive Development,* ed. J. Kaufman and J. Baer. Cambridge: Cambridge University Press.

Potts, R. 1996. *Humanity's Descent: The Consequences of Ecological Instability.* New York: Avon.

Potts, R. 1998. Variability selection in hominid evolution. *Evolutionary Anthropology* 7 (3):81–96.

Povinelli, D., with J. Reaux, L. Theall, and S. Giambrone. 2000. *Folk Physics for Apes: The Chimpanzee's Theory of How the World Works.* Oxford: Oxford University Press.

Powell, A., S. Shennan, and M. Thomas. 2009. Late Pleistocene demography and the appearance of modern human behavior. *Science* 324:298–1301.

Prinz, J. 2009. *The Emotional Construction of Morals.* Oxford: Oxford University Press.

Prinz, J. 2010. Against moral nativism. In *Stich and His Critics,* ed. D. Murphy and M. Bishop. Oxford: Blackwell.

Pyne, S. 1998. *Burning Bush: A Fire History of Australia.* Seattle: University of Washington Press.

Read, D. 2006. Tasmanian knowledge and skill: Maladaptive imitation or adequate technology? *American Antiquity* 71 (1):164–184.

Reader, S., and K. Laland. 2002. Social intelligence, innovation, and enhanced brain size in primates. *Proceedings of the National Academy of Sciences of the United States of America* 99 (7):4436–4441.

Richerson, P. Forthcoming. Rethinking paleoanthropology: A world queerer than we had supposed. In *The Evolution of Mind,* ed. G. Hatfield. Philadelphia: University of Pennsylvania Press.

Richerson, P., and R. Boyd. 1999. Complex societies: The evolutionary origins of a crude superorganism. *Human Nature* 10 (3):253–289.

Richerson, P., and R. Boyd. 2001. The evolution of subjective commitment to groups: A tribal instinct hypothesis. In *The Evolution of Commitment*, ed. R. M. Nesse, 186–220. New York: Russell Sage Foundation.

Richerson, P., and R. Boyd. 2002. Institutional evolution in the Holocene: The rise of complex societies. In *The Origins of Human Social Institutions*, ed. W. G. Runciman, 197–234. London: British Academy.

Richerson, P., and R. Boyd. 2005. *Not by Genes Alone: How Culture Transformed Human Evolution*. Chicago: University of Chicago Press.

Richerson, P., R. Boyd, and R. Bettinger. 2001. Was agriculture impossible during the Pleistocene but mandatory during the Holocene? A climate change hypothesis. *American Antiquity* 66:387–411.

Ricklefs, R. E., and D. Schulter, eds. 1993. *Species Diversity in Ecological Communities*. Chicago: Chicago University Press.

Robson, A., and H. Kaplan. 2003. The evolution of human life expectancy and intelligence in hunter-gatherer economies. *American Economic Review* 93 (1):150–169.

Ross, D. 2006. Evolutionary game theory and the normative theory of institutional design: Binmore and behavioral economics. *Politics, Philosophy, and Economics* 5 (1):51–80.

Sapolsky, R. 2002. *A Primate's Memoir*. New York: Touchstone Books.

Saunders, S. 2009. Costly signalling: A work in progress. *Biology and Philosophy* 24 (3):405–416.

Schelling, T. 2001. Commitment: Deliberate versus involuntary. In *Evolution and the Capacity for Commitment*, ed. R. Nesse, 49–56. New York: Russell Sage Foundation.

Schelling, T. C. 2006. *Strategies of Commitment and Other Essays*. Cambridge, MA: Harvard University Press.

Schlegel, A., and H. Barry. 1980. The evolutionary significance of adolescent initiation ceremonies. *American Ethnologist* 7 (4):696–715.

Scott, T. 2010. The evolution of moral cognition. Doctoral Dissertation, Department of Philosophy, Victoria University of Wellington.

Seabright, P. 2006. The evolution of fairness norms: An essay on Ken Binmore's *Natural Justice*. *Politics, Philosophy, and Economics* 5 (1):33–50.

Seabright, P. 2008. Warfare and the multiple adoption of agriculture after the last Ice Age. IDEI Working Paper, n. 522, April.

Seabright, P. 2010. *The Company of Strangers: A Natural History of Economic Life*. Princeton: Princeton University Press.

Searcy, W., and S. Nowicki. 2005. *The Evolution of Animal Communication: Reliability and Deception in Signaling Systems*. Princeton: Princeton University Press.

Shaw-Williams, K. 2011. Trackway reading and the hominin narrative faculty. Master's Thesis, Department of Philosophy, Victoria University of Wellington.

Shea, J. 2003. Neandertals, competition, and the origins of modern human behavior in the Levant. *Evolutionary Anthropology* 12:173–187.

Shea, J. 2006. Child's play: Reflections on the invisibility of children in the Paleolithic record. *Evolutionary Anthropology* 15:212–216.

Shea, J. 2009. The impact of projectile weaponry on Late Pleistocene hominin evolution. In *The Evolution of Hominid Diets*, ed. J. J. Hublin and M. P. Richards, 187–198. Berlin: Springer Science.

Shennan, S. 2002. *Genes, Memes, and Human History: Darwinian Archaeology and Cultural Evolution*. London: Thames and Hudson.

Shultziner, D., T. Stevens, M. Stevens, B. A. Stewart, R. Hannagan, and G. Saltini-Semerari. 2010. The causes and scope of political egalitarianism during the Last Glacial: A multi-disciplinary perspective. *Biology and Philosophy* 25 (3):319–346.

Skyrms, B. 1996. *The Evolution of the Social Contract*. Cambridge: Cambridge University Press.

Skyrms, B. 2003. *The Stag Hunt and the Evolution of Social Structure*. Cambridge: Cambridge University Press.

Skyrms, B. 2010. *Signals: Evolution, Learning, and Information*. Oxford: Oxford University Press.

Smetana, J. G. 1989. Toddlers' social interactions in the context of moral and conventional transgressions in the home. *Developmental Psychology* 25:499–508.

Smetana, J. G. 1999. The role of parents in moral development: A social domain analysis. *Journal of Moral Education* 28:311–321.

Smith, E. A., and R. Bliege Bird. 2005. Costly signaling and cooperative behavior. In *Moral Sentiments and Material Interests: The Foundations of Cooperation in Economic Life*, ed. H. Gintis, S. Bowles, R. Boyd, and E. Fehr, 115–148. Cambridge, MA: MIT Press.

Smith, E. A., R. Bliege Bird, and D. Bird. 2003. The benefits of costly signaling: Meriam turtle hunters. *Behavioral Ecology* 14 (1):116–126.

Sober, E., and D. S. Wilson. 1998. *Unto Others: The Evolution and Psychology of Unselfish Behavior*. Cambridge, MA: Harvard University Press.

Soltis, J., R. Boyd, and P. Richerson. 1995. Can group-functional behaviors evolve by cultural group selection? An empirical test. *Current Anthropology* 63:473–494.

Sosis, R., H. Kress, and J. S. Boster. 2007. Scars for war: Evaluating alternative signaling explanations for cross-cultural variance in ritual costs. *Evolution and Human Behavior* 28:234–247.

Sperber, D. 1996. *Explaining Culture: A Naturalistic Approach.* Oxford: Blackwell.

Sperber, D. 1997. Intuitive and reflective beliefs. *Mind and Language* 12:67–83.

Sperber, D. 2000. Metarepresentations in an evolutionary perspective. In *Metarepresentations: A Multidisciplinary Perspective,* ed. D. Sperber, 117–137. Oxford: Oxford University Press.

Sperber, D. 2001. An evolutionary perspective on testimony and argumentation. *Philosophical Topics* 29:401–413.

Sperber, D., F. Clément, C. Heintz, O. Mascaro, H. Mercier, G. Origgi, and D. Wilson. 2010. Epistemic vigilance. *Mind and Language* 25 (4):359–393.

Sripada, C. and S. Stich. 2006. A framework for the psychology of norms. In *The Innate Mind: Culture and Cognition,* ed. P. Carruthers, S. Laurence, and S. Stich. New York: Oxford University Press.

Stanford, C. 1999. *The Hunting Ape: Meat Eating and the Origins of Human Behavior.* Princeton: Princeton University Press.

Stanovich, K. 2004. *The Robot's Rebellion.* Chicago: University of Chicago Press.

Sterelny, K. 2003. *Thought in a Hostile World.* New York: Blackwell.

Sterelny, K. 2006a. Local ecological communities. *Philosophy of Science* 73 (2):215–231.

Sterelny, K. 2006b. The evolution and evolvability of culture. *Mind and Language* 21 (2):137–165.

Sterelny, K. 2007. Social intelligence, human intelligence, and niche construction. *Philosophical Transactions of the Royal Society of London, Series B: Biological Sciences* 362 (1480):719–730.

Sterelny, K. 2010. Minds: Extended or scaffolded. *Phenomenology and the Cognitive Sciences* 9:65–481.

Sterelny, K. 2011. From hominins to humans: How Sapiens became behaviourally modern. *Philosophical Transactions of the Royal Society of London, Series B: Biological Sciences* 366:809–822.

Stich, S. 1993. Moral philosophy and mental representation. In *The Origin of Values,* ed. M. Hechter, L. Nadel, and R. Michod, 215–228. New York: Aldine de Gruyer.

Stiner, M. C. 2001. Thirty years on: The "Broad Spectrum Revolution" and Paleolithic demography. *Proceedings of the National Academy of Sciences of the United States of America* 98 (13):6993–6996.

Stiner, M. C. 2002. Carnivory, coevolution, and the geographic spread of the genus Homo. *Journal of Archaeological Research* 10 (1):1–63.

Stiner, M. C., and S. L. Kuhn. 2006. Changes in the "connectedness" and resilience of Paleolithic societies in Mediterranean ecosystems. *Human Ecology* 34 (5):693–712.

Stout, D. 2002. Skill and cognition in stone tool production: An ethnographic case study from Irian Jaya. *Current Anthropology* 43 (5):693–722.

Stout, D. 2011. Stone toolmaking and the evolution of human culture and cognition. *Philosophical Transactions of the Royal Society of London, Series B: Biological Sciences* 366:1050–1059.

Takezawa, M., M. Gummerum, and M. Keller. 2006. A stage for the rational tail of the emotional dog: Roles of moral reasoning in group decision making. *Journal of Economic Psychology* 27:117–139.

Tattersall, I. 2009. Human origins: Out of Africa. *Proceedings of the National Academy of Sciences of the United States of America* 106 (38):16018–16021.

Tehrani, J., and F. Riede. 2008. Towards an archaeology of pedagogy: Learning, teaching, and the generation of material culture traditions. *World Archaeology* 40 (3):316–331.

Tennie, C., J. Call, and M. Tomasello. 2009. Ratcheting up the ratchet: On the evolution of cumulative culture. *Philosophical Transactions of the Royal Society of London, Series B: Biological Sciences* 364:2405–2415.

Thieme, H. 1997. Lower Palaeolithic hunting spears from Germany. *Nature* 385: 807–811.

Thornton, A., and N. J. Raihani. 2008. The evolution of teaching. *Animal Behaviour* 75:1823–1836.

Tomasello, M. 1999a. *The Cultural Origins of Human Cognition*. Cambridge, MA: Harvard University Press.

Tomasello, M. 1999b. The human adaptation for culture. *Annual Review of Anthropology* 28:509–529.

Tomasello, M. 2003. *Constructing a Language: A Usage-Based Theory of Language Acquisition*. Cambridge, MA: Harvard University Press.

Tomasello, M. 2008. *Origins of Human Communication*. Cambridge, MA: MIT Press.

Tomasello, M. 2009. *Why We Cooperate*. Cambridge, MA: MIT Press.

Tomasello, M., and M. Carpenter. 2007. Shared intentionality. *Developmental Science* 10 (1):121–125.

Tomasello, M., M. Carpenter, J. Call, T. Behne, and H. Moll. 2005. Understanding and sharing intentions: The origins of cultural cognition. *Behavioral and Brain Sciences* 28:675–691.

Tooby, J., and L. Cosmides. 1992. The psychological foundations of culture. In *The Adapted Mind*, ed. J. Barkow, L. Cosmides, and J. Tooby, 19–136. Oxford: Oxford University Press.

Varki, A., D. H. Geschwind, and E. E. Eichler. 2008. Explaining human uniqueness: Genome interactions with environment, behaviour, and culture. *Nature Reviews: Genetics* 9 (10):749–763.

Visalberghi, E., and J. Anderson. 2008. Fair game for chimpanzees. *Science* 319: 282–283.

Wadley, L. 2001. What is cultural modernity? A general view and a South African perspective from Rose Cottage Cave. *Cambridge Archaeological Journal* 11 (2):201–221.

Wadley, L. 2010. Compound-adhesive manufacture as a behavioural proxy for complex cognition in the Middle Stone Age. *Current Anthropology* 51 (Supplement 1):S111–S119.

Walker, R., K. Hill, H. Kaplan, and G. McMillan. 2002. Age-dependency in hunting ability among the Ache of Eastern Paraguay. *Journal of Human Evolution* 42 (6):639–657.

Warneken, F. Forthcoming. The origins of human cooperation from a developmental and comparative perspective. In *The Evolution of Mind*, ed. G. Hatfield. Philadelphia: University of Pennsylvania Press.

Warneken, F., B. Hare, A. Melis, D. Hanus, and M. Tomasello. 2007. Spontaneous altruism by chimpanzees and young children. *PLOS Biology* 5(7):e184.

Warneken, F., and M. Tomasello. 2006. Altruistic helping in human infants and young chimpanzees. *Science* 311:1301–1303.

Warneken, F., and M. Tomasello. 2009. Varieties of altruism in children and chimpanzees. *Trends in Cognitive Sciences* 13 (9):397–402.

Washburn, S. L., and C. Lancaster. 1968. The evolution of hunting. In *Man the Hunter*, ed. R. B. Lee and I. DeVore, 293–303. Chicago: Aldine.

Weber, B., and D. Depew. 2003. *Evolution and Learning: The Baldwin Effect Reconsidered.* Cambridge, MA: MIT Press.

Weisdorf, J. 2005. From foraging to farming: Explaining the Neolithic revolution. *Journal of Economic Surveys* 19 (4):561–586.

West, S. A., A. S. Griffin, and A. Gardner. 2007. Social semantics: Altruism, cooperation, mutualism, strong reciprocity, and group selection. *Journal of Evolutionary Biology* 20 (2):415–432.

West-Eberhard, M. J. 2003. *Developmental Plasticity and Evolution*. Oxford: Oxford University Press.

Whiten, A. 2001. The scope of culture in chimpanzees, humans, and ancestral apes. *Philosophical Transactions of the Royal Society of London, Series B: Biological Sciences* 366:997–1007.

Whiten, A. 2005. The second inheritance system of chimpanzees and humans. *Nature* 437:52–55.

Whiten, A., K. Schick, and N. Toth. 2009. The evolution and cultural transmission of percussive technology: Integrating evidence from palaeoanthropology and primatology. *Journal of Human Evolution* 30:1–16.

Williams, G. C. 1966. *Adaptation and Natural Selection*. Princeton: Princeton University Press.

Wrangham, R. 1999. Evolution of coalitionary killing. *Yearbook of Physical Anthropology* 42:1–30.

Wrangham, R. 2009. *Catching Fire: How Cooking Made Us Human*. London: Profile Books.

Wynn, T., and F. Coolidge. 2004. The expert Neanderthal mind. *Journal of Human Evolution* 46:467–487.

Zahavi, A., and A. Zahavi. 1997. *The Handicap Principle: A Missing Piece of Darwin's Puzzle*. Oxford: Oxford University Press.

Zilhão, J. 2007. The emergence of ornaments and art: An archaeological perspective on the origins of "behavioural modernity." *Journal of Archaeological Research* 15:1–54.

Zilhão, J., D. Angelucci, E. Badal-García, F. d'Errico, F. Daniel, L. Dayet, K. Douka, et al. 2010. Symbolic use of marine shells and mineral pigments by Iberian Neandertals. *Proceedings of the National Academy of Science* 107:1023–1028.

# Index